Geological Objects and Structures in 3D

Geological Objects and Structures in 3D

Observation, Interpretation and Building of 3D Models

Dominique Frizon de Lamotte,
Pascale Leturmy, Pauline Souloumiac
and Adrien Frizon de Lamotte

CRC Press is an imprint of the
Taylor & Francis Group, an **informa** business

Cover photo: 'Layered rocks on the southern coast of Crete' © zavgsg – Fotolia.com

CRC Press/Balkema is an imprint of the Taylor & Francis Group, an informa business

Typeset by Apex CoVantage, LLC

Originally published in France as:
Objets et structures géologiques en trois dimensions. Observation, interprétation et construction: Observations, interprétation et construction de modèles en 3D

By Dominique FRIZON DE LAMOTTE, Pascale LETURMY, Pauline SOULOUMIAC & Adrien FRIZON DE LAMOTTE

Library of Congress Cataloging-in-Publication Data
Names: Frizon de Lamotte, Dominique, author. | Leturmy, P. (Pascale), author. | Souloumiac, Pauline, author. | Frizon de Lamotte, Adrien, author.
Title: Geological objects and structures in 3D / by Dominique Frizon de Lamotte, Pascale Leturmy, Pauline Souloumiac, Adrien Frizon de Lamotte.
Other titles: Geological objects and structures in three D
Description: Boca Raton : CRC Press, 2020. | Includes bibliographical references and index.
Identifiers: LCCN 2020014737 (print) | LCCN 2020014738 (ebook) | ISBN 9780367497330 (hbk) | ISBN 9780367497507 (paperback) | ISBN 9781003047230 (ebook)
Subjects: LCSH: Geology—Maps.
Classification: LCC QE36 .F79 2020 (print) | LCC QE36 (ebook) | DDC 912.01/4023551—dc23
LC record available at https://lccn.loc.gov/2020014737
LC ebook record available at https://lccn.loc.gov/2020014738

Published by: CRC Press/Balkema
Schipholweg 107C, 2316 XC Leiden, The Netherlands
e-mail: Pub.NL@taylorandfrancis.com
www.routledge.com – www.taylorandfrancis.com

ISBN: 978-0-367-49733-0 (Hbk)
ISBN: 978-0-367-49750-7 (Pbk)
ISBN: 978-1-003-04723-0 (eBook)
DOI: 10.1201/9781003047230
DOI: https://doi.org/10.1201/9781003047230

Contents

Preface

> The best method of getting a grasp of structural or tectonic geology is to attempt the construction of a geological map . . . there are few more engrossing or interesting pursuits than that of unravelling geological structure.
>
> (A. Geikie, 1912)

Understanding the relationships between strata forms the basis for much geological knowledge – and these patterns are displayed on geological maps. Dominique Frizon de Lamotte and colleagues from the CY Cergy Paris Université, on the outskirts of Paris, have put together this lavishly illustrated manual to assist in these endeavors. As they note, the observation, interpretation and construction of 3D models lies at the heart of applied geology and has driven many of the great discoveries in the science. This book is about structural geology – especially interpreting the geometry of layered strata where they become folded and faulted. The starting point is the geological map, and the challenge – understanding and communicating this understanding to others – has been at the forefront of the skill set required by geologists since the birth of the science.

Taking observations from the earth's surface and forecasting what happens in the subsurface involves stepping into the unknown. It is a mission that has inspired great works – both of science and of fiction. But it's not just about satisfying curiosity; understanding the subsurface lies at the heart of applied geology, exploring and extracting earth resources, be they minerals, hydrocarbons or water demands predicting where these lie at depth – so as not to waste money by drilling in the wrong places. And it's not just about money; drilling into the wrong things, such as otherwise unidentified pockets of over-pressured fluids, can have catastrophic consequences both to the engineers on the surface doing the drilling and the environment. Likewise, geological formations are increasingly being sought to host the waste products of industrial processes, especially from the energy sector. This includes pumping CO_2 into former hydrocarbon reservoirs as well as engineering solutions for storing radioactive materials associated with the nuclear industries. Indeed, all big civil engineering projects demand geological knowledge of the subsurface – be they excavating the Channel Tunnel, the Crossrail beneath London or the West Island Line and Express Rail Link beneath Hong Kong. Assessing many natural hazards demands geological understanding of the subsurface; for example, knowing the architecture of geological faults is important for assessing earthquake hazard in tectonically active parts of the world.

Building knowledge of the subsurface involves exercises in deductive reasoning. In *Objects and Geological Structures in 3D*, Dominique Frizon de Lamotte and colleagues take readers systematically through an increasingly elaborate collection of reasonings and methods. Though entirely modern in approach and illustration, the narrative is essentially one of the history of our science and its use in developing geological knowledge for the benefit of society. This all stems from ideas developed centuries ago, starting with the Danish philosopher-scientist Nicolas Steno in the 17th century. He realized that successions of sedimentary rocks built up one layer on top of another so that, by recognizing superposition, the observer can deduce a sequence in geological time. He also established that, by identifying cross-cutting relationships, a history of igneous intrusions could be established. The general importance was that histories of events in the geological past could be deduced from observations of rock relationships made today. Just how long these histories might be began to be realized in the late 18th century by the Scot James Hutton. He recognized that rock successions contained unconformities which represented long periods of time during which strata could be lithified by burial, deformed into folds, tilted, then eroded, subsided and then buried again by further layers of strata. Unconformities remain the fundamental markers for dividing up rock successions and tracing them across entire sedimentary basins.

While through the 19th century pioneers of geology added to philosophical discussions of the earth and humankind's place in it, the new science also played a pivotal role in resourcing the Industrial Revolution. So many of the basic methods and approaches that now inform knowledge of the structure of the earth have arisen through applying geological science to the exploitation of earth resources. This began as the 18th turned into the 19th century with the British geologist William Smith and others, who created the earliest detailed geological maps of large regions. What began as essentially spatial catalogs of different strata and the mineral resources they hosted quickly evolved into tools for forecasting the near subsurface. This in turn led to a suite of technical textbooks aimed at taking surface geology from geological maps into the subsurface, as illustrated on cross-sections. Much of this was aimed at assisting deep mining operations, especially for coal, in the industrializing countries of Europe and the United States. These endeavors generally involved forecasting the continuity of layers, including their offsets across faults, for relatively short distances away from direct observations at the earth's surface or in a mine gallery. Consequently, the methods used by these texts are generally based on simple trigonometry, for example, estimating the thickness of a coal seam based on its outcrop pattern and the measured inclination of strata.

The quote at the start of this Preface comes from James Geikie's textbook on *Structural and Field Geology*, which typified the extent of approaches used to interpret the subsurface. But as the 20th century progressed, society's constant demand for cheap energy through the search for oil and gas forced geologists to confront greater interpretational challenges. Exploitable oil and gas lie at significantly greater depths below the earth's surface than coal mines, so extrapolation of geological observations had to be extrapolated correspondingly further. With this came the push into increasingly complex structures – such as those on the edge of young mountain ranges. Along with the trigonometric methods came physical theories and the search for comparative structures either identified in outcrops or on cross-sections drawn from other parts of the world, together with the results of deformation experiments performed on analog materials such as layers of sand and modeling clay.

Many new approaches were pioneered in the foothills of the Canadian Cordillera. In the 1930s and 1940s, Ted Link, a leading geologist with the Imperial Oil company (part of Exxon), was one of the first to compare his interpretations of complex fold and structures on cross-sections with models he created by squashing layers of sand and clay in a vice. Subsequently, cross-sections through the Canadian foothills were refined through seismic imaging.

But there still remained uncertainties in how subsurface structure was interpreted – there were many ways of drawing cross-sections through prospective oil and gas fields. To narrow the options, in the 1960s the community of exploration geologists based in Calgary developed protocols for drawing cross-sections that could, by following a narrow set of geometric rules (concentric folding, where layers retain constant thickness during deformation, and simple faulting), be unraveled to display the pre-deformation arrangement of strata. These illustrations are called balanced cross-sections. To be acceptable, the cross-section had to balance – the strata unraveled onto a restored section to have equal undeformed lengths without gaps and overlaps. If the structural geometry on the cross-section couldn't be unraveled, then this cross-section and the structural geometry it illustrated would be discarded. The geometric rules for cross-section construction were formalized in the late 1960s by Clint Dahlstrom of the Chevron oil company and led directly to codified versions that quantified the geometry of folds and explicitly defined fault geometries. This in turn led, in the 1980s and 1990s, to a series of computer programs that assisted cross-section construction and structural restoration. They continue to be developed into the 2020s. Computer-assisted structural interpretation is the norm in the oil and gas industry and drives much academic research too.

While Dahlstrom's rules continue to underpin subsurface interpretation in thrust belts some 50 years after their proposition, some examples exhibit rather more complex geology. And strata can be deformed by processes that do not simply relate to plate tectonics. This is exemplified by the behavior of thick salt deposits. Salt is mobile – having a viscosity much lower than other rocks in the upper crust – and it has a low density, making it more buoyant than other sedimentary rocks in the subsurface. The result is mobility – salt flows, and so the surrounding strata (and earth's surface) folds in response. In the past two decades, understanding of the complex structural and stratigraphic relationships associated with salt mobility has changed the way we understand many sedimentary basins. This has led to important discoveries of oil and gas accumulations, especially along parts of the world's continental margins. The salt (not just halite but also the stringers of potash) is economically important in itself, and old salt mines are proposed as sites for the long-term storage of nuclear waste. Geologists need excellent skills in structural geology to support these engineering endeavors, to forecast the structure and future mobility of the salt.

Looking forward, the subsurface will remain important to society – geologists will need the skills to create interpretations – increasingly in structurally complex regions. But it's all very well for a geologist to be able to visualize complex structures and create geometrically reasonable three-dimensional interpretations. The next challenge is to be able to communicate these insights to non-specialists. For applied geologists, effective communication is imperative because other people are likely to be making engineering and financial decisions on the basis of these interpretations. Traditionally geologists might have tried to draw perspective views or construct physical cross-sections that can be built into a gridwork – relying on their abilities to sketch using paper and pencil. Nowadays geologists use computer graphics, maybe created physically through 3D printing, perhaps accompanied by an animation depicting structural evolution, constructing interpretations using strict geometric rules and quantified structural geometries.

This book builds up from basic foundations, setting up the layered structure of the earth and covering basic visualizations of dipping strata and their relationship to topography. It then looks at unconformities, with examples on maps, landscapes and cross-sections, and includes a section on using stratigraphic relationships to date structures. With these stratigraphic concepts in place, the next topics are concerned with faults and faulting. Again, field examples come to the fore, but they are set up from a mechanical basis using Mohr circles and relationships between fault slip and fault propagation. Next up come folds, specifically those developed above regional detachment surfaces, such as happens in the Jura hills on the edge of the Alps. These concepts lead in turn to the relationships between folds and thrust faults, showing how these relationships can be quantified geometrically. The new chapter on salt tectonics follows up the "classical" fold and thrust concepts. The whole thing is wrapped up with illustrations of 3D model building, illustrated by stratigraphic relationships and crystal forms. It all goes to show how three-dimensional understanding is fundamental to successful geological study.

So *Objects and Geological Structures in 3D* is about structural geology – building interpretations of faults and folds in layered strata. The focus is on structures formed on the margins of mountain belts, so is in Dahlstrom's lineage. Examples from North America, which would have been familiar to Dahlstrom and his colleagues in Calgary, feature here. But most of the structural examples chosen by Dominique Frizon de Lamotte and colleagues come from the Tethyan chains of Europe, North Africa and the Middle East. This befits a book that was originally published in French and is now receiving its English translation. This English edition has been updated to include salt tectonics, increasing the book's value in supporting interpretation of a greater range of subsurface structures. The diversity of examples is a real strength, for if the reader is seeking to find examples of particular structures then it is good to draw a wide net – to appreciate variety. The beautiful illustrations of 3D block diagrams coupled with photographs of natural structures should inspire undergraduate and postgraduate students as well as amateur geologists curious about how the earth is structured. There are indeed fewer more engrossing pursuits.

Rob Butler
Aberdeen, January 2020

Foreword

Geology is a scientific discipline where a 3D view is important – even essential. When starting to learn geology, as a first exercise students should be able to gain a 3D vision of geological maps, which like all maps are 2D objects, and interpret them.

Many people have an objective difficulty in "seeing in 3D," that is, in achieving a mental representation of a dimension that is not shown. To help them in this task we dissect the geometry of geological objects and structures. We use simple mathematical (mainly trigonometric) notions and the mechanics of continuous media, but this is a geology book illustrated by many colored diagrams and field photographs. The notions presented are simple and do not give rise to any conceptual difficulties. Nevertheless, we wanted to show that it is possible to learn a great deal from a good understanding of 3D geometries. By looking into things more deeply in this way, we are sometimes led to a number of counter-intuitive ideas, such as, for example, in the formation of folds.

The original feature of this book is that it offers a wide range of objects, which anyone can use or make in line with an educational approach that combines digital creation and object manipulation. In fact, our computer-designed prototypes are saved in a format from which they can be printed in 3D. Three types of objects are presented: (1) models which help to see things in 3D and thus understand particular structures, (2) models where the third dimension offers an approach to successive geometries (kinematics) during the formation of particular geological structures, and (3) models that provide the opportunity to move different parts relative to each other to generate structures like faults.

We venture that through reading this book and using our models, and possibly creating other objects by themselves, students will be helped to find their way in this 3D world, which is often confusing at first sight. The target audience is students from first degree to Master's level, trainee teachers, secondary school science teachers and amateur or professional geologists. We also want to reach the growing network of "fablabs," whether or not they are university-based.

The basis of this book is the creation and production of teaching aids. The step from one to the other requires the use of computer-aided design and a knowledge of 3D printing procedures. It is therefore essentially a collaborative work that has involved our students from degree to doctorate level and our colleagues from the Geosciences and Environment Department of CY Cergy Paris Université. It will become apparent from reading the book that our research work has nourished our approach to teaching, as is appropriate in an academic context.

The book is divided into six chapters, each dealing with a particular topic: the mapping of geological strata, unconformities, faults, detached folds, folds associated with faults and finally salt tectonics. The illustrations are mainly from our own fieldwork or that of colleagues who have kindly allowed us to use them. In this respect, we would like to thank in particular Jean-Claude Ringenbach, an expert structural geologist with Total, who has allowed us to draw on his rich photo library.

The photos of the prototypes were taken by Alexis Chézière© for CY Cergy Paris Université, which has given permission for them to be used. Technical terms are explained in a glossary at the back of the book. A stratigraphic chart is included (after the glossary) for use when reading to include the aspect of geological time. The book also provides practical tips and recommendations for 3D printing, as well as some ideas for future developments.

Chapter 1

Geology, a story of layers

> In cartography, the insoluble problem of *projections* arises from the impossibility of representing on a plane, without deforming it, a curved surface.
>
> (J. Gracq, 1986)

1.1 The globe and the various layers that make it up

The upper part of the globe, the so-called lithosphere, is formed by a series of layers (Fig. 1.1). Thus, we distinguish from bottom to top:

- the upper (lithospheric) mantle consisting of peridotites; and
- the continental or oceanic crust.

The continental crust, of general granitic composition, consists of two layers (three for some authors) with different mechanical behaviors. The upper crust exhibits a rather brittle behavior while the lower crust holds ductile behavior. As for the oceanic crust, it is also formed of layers that are of different natures: peridotite below, then gabbro and finally basalt at the top. The deeper layers (lower mantle and core) are inaccessible to direct observation. We will not talk about them in this book, but they are recalled for the record in Figure 1.1.

At the top of the edifice, the sedimentary cover is a stack of layers (in the sedimentary domain, we also speak of strata) of varied nature (limestone, marl, clay, sand, sandstone, salt etc.) and contrasting mechanical behavior. In this book, we are interested primarily, but not exclusively, in this very superior part of the system that is at the interface with the biosphere.

These layers that form the sedimentary basins can be folded and/or faulted. They can also be integrated into the mountain ranges, with in this case much larger deformations. When the mountain ranges are young or even still active, considerable reliefs (Andes and Himalayas) mark them. When these chains are old, they are integrated into the continental crust itself and form what is called the "basement." The basement is separated from the sedimentary cover by a major discontinuity called unconformity (for developments on the concepts of basement and sedimentary cover, see Chapter 2). The basement can be at the surface in continental areas without sedimentary cover. The areas where the basement is exposed represents about 30% of the surface of the continental shelf. They are called "old massif," "shield" or, if the basement is very old, "craton." Basement and cover have, in general, very different mechanical behaviors. As a result, the basement-cover interface is frequently the site of significant decoupling generating different deformation styles on both sides.

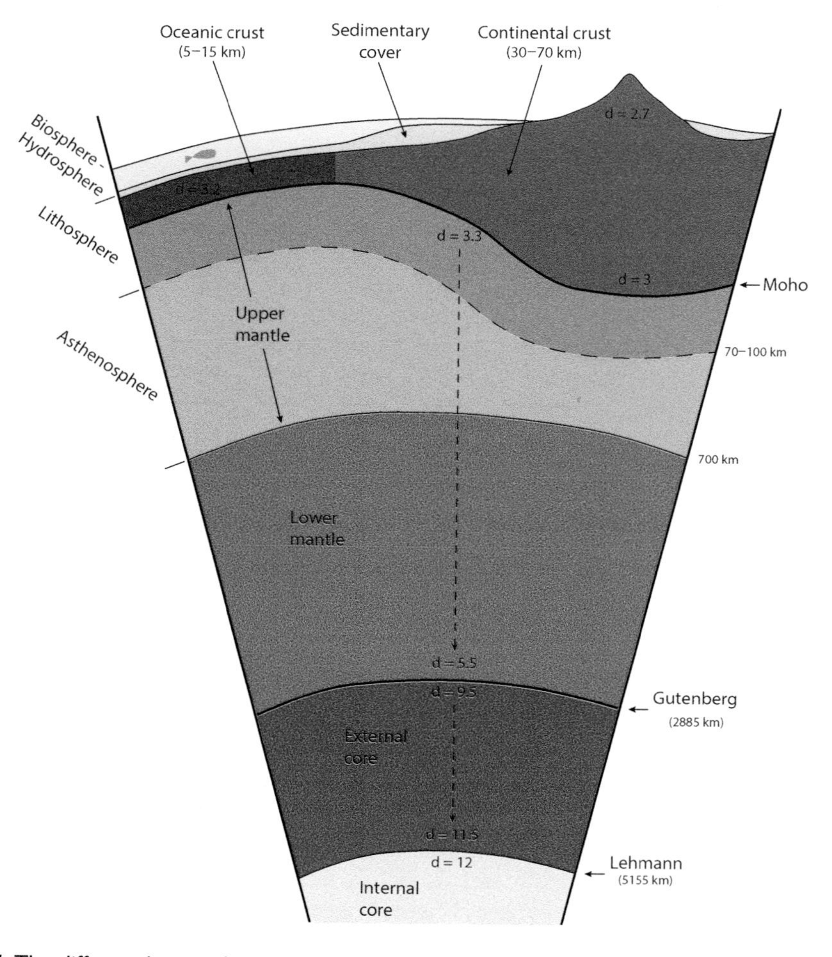

Figure 1.1 The different layers of the earth system.

1.2 Strike and dip of a layer

The dip of a geological layer is the angle between a horizontal plane and the plane of the layer (β in Fig. 1.2), the intersection between the two planes being the strike or azimuth (α). The measurement of a dip obviously does not concern only the sedimentary layers; we can measure the dip of any plane having a geological significance: fault plane, the wall of a vein, the contact with an intrusion etc. A dip is measured in a vertical plane passing through the line of greatest slope of the plane to be measured. To be perfectly located in geographical space, we must also know the strike (α), that is to say, the angle measured in the horizontal plane between a horizontal line contained in the geological layer and the geographical north (Fig. 1.2).

In practice, there are two methods for measuring the dip of a layer in the field. The first one, called "the French method," consists in measuring initially the strike of the horizontal

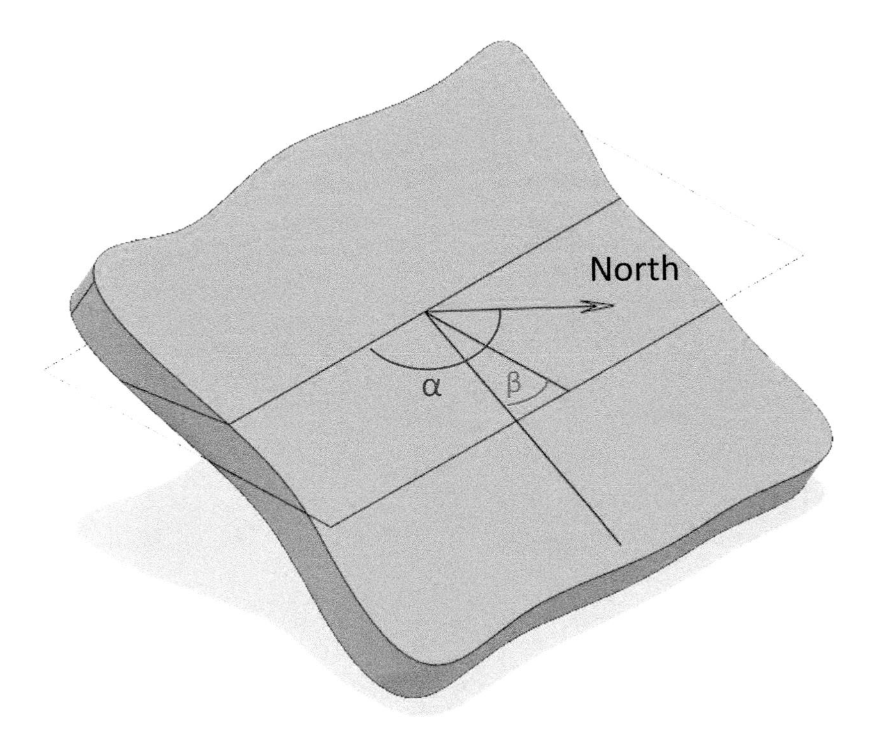

Figure 1.2 Notion of dip of a layer (see explanations in the text).

plane (α in Fig. 1.2). This angle α is measured between 0° and 180° in the east dial of the compass. The plane is written in the following manner N x° E ($0 \leq x \leq 180$), meaning: the horizontal has an azimuth of x° measured in the east dial of the compass. Measuring in this dial is a pure convention. The dip is then measured ($0 \leq \beta \leq 90$) using a clinometer. Finally, it is necessary to indicate the direction of the dip. There remains indeed an indeterminacy because, for a given strike and dip, there are two directions of dip possible. Thus, the final data will be written, for example, N 45° E 60° SW, meaning: an azimuth plane 45° E and a dip 60° to the SW. The so-called "English method" is, let us admit it, simpler and more effective. It consists in giving the angular value of the dip ($0 \leq \beta \leq 90$) then its direction, thanks to a single measurement between 0° and 360°. The final data will therefore be written for the previous example: 60° N 135°. If the dip is in the other direction, using the "French method" N 45 E 60° NE, then the data would be written using the "English method": 60° N 225°.

The dip of a layer is measured in the plane containing the line of greatest slope. This can and should be verified when you have access to 3D geometry. Where we have access only to a section (2D by definition), which is frequently the case, we do not know, *a priori*, if we are dealing with a "real dip" (that is to say, measured in the plane containing the line of greatest slope) or with an "apparent dip" (that is, measured in another vertical plane). The apparent dip is necessarily lower than the actual dip. Finally, if the measurement plane contains the horizontal of the place, the apparent dip will be zero.

1.3 Geometric relationships between dip and topography using the three points method

In the general case, the geologist derives their rationale from a geological map. On this map, the geological layers, represented by colors corresponding to geological ages (see Stratigraphic chart), interfere with the topography represented by contour lines. The map gives us direct access to the thickness of a layer only in the particular case where this layer is vertical (Fig. 1.3).

In the general case, one only has access to the outcropping width (L) of a layer that depends on its thickness (e) and the angular value of the slope (θ).

A simple way to comprehend this is to look at the frequent example of horizontal layers forming plateaus intersected by valleys (Fig. 1.4). In a general manner, we can write:

$$L = e / \tan\theta$$

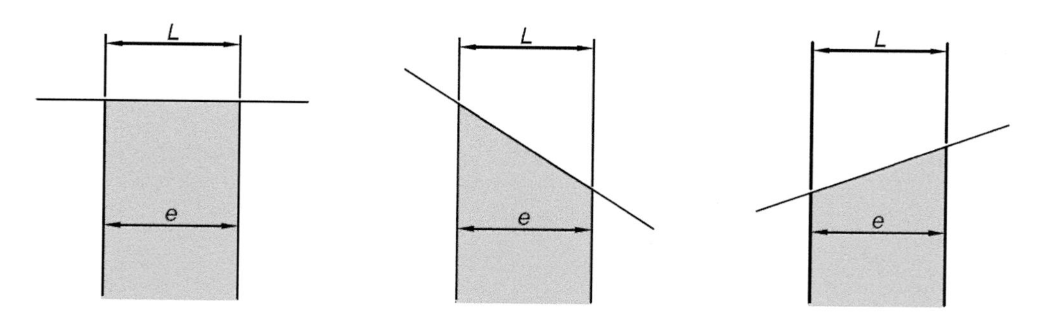

Figure 1.3 Series of three sections showing that for a vertical layer the outcropping width (L) is always equal to the thickness of the layer (e). The fine line represents the topographic surface.

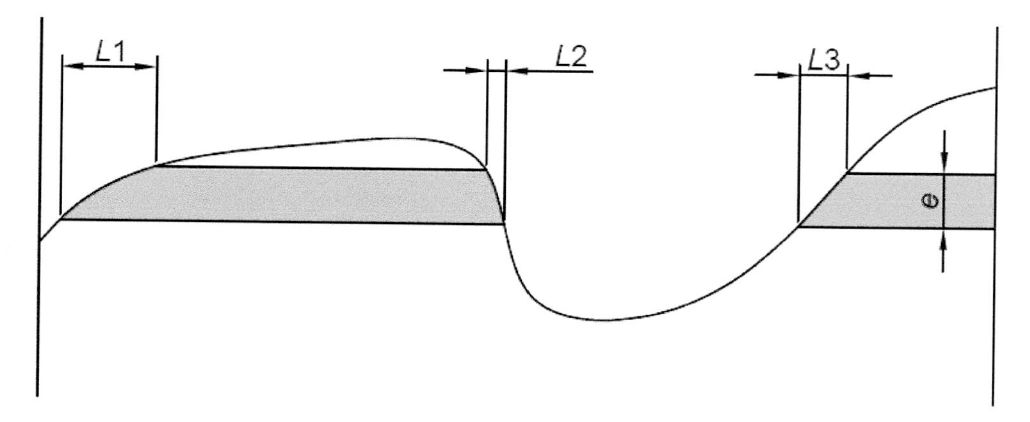

Figure 1.4 Schematic section showing that for a horizontal layer of constant thickness (e) the width at the outcropping (L) will vary according to the angle of the slope (θ) such that: $L = e / \tan\theta$.

The width in the outcrop is therefore, in this case, proportional to the thickness of the layer (e) and inversely proportional to the slope (θ). For a given thickness, the outcropping width depends only on the topographic slope:

$$e = L1 \tan\theta = L2 \tan\theta' = L3 \tan\theta'$$

Let us now examine the case of the intersection of any dip layer (β) with a slope of a given angle (θ) (Fig. 1.5). We have thus:

$$e = L' \sin(\theta + \beta) \rightarrow L' = e / \sin(\theta + \beta)$$

$$L = L' \cos\theta \rightarrow L = e \frac{\cos\theta}{\sin(\theta + \beta)} \rightarrow e = L \frac{\sin(\theta + \beta)}{\cos\theta}$$

In a practical way, a simple method, called the three points method, allows the simple calculation of the dip of a geological layer from the examination of the modalities of intersection between the topography and the geological layers in a valley. We choose three points located on the same boundary layer because two of them are on the same contour line and the third at the bottom of the valley, that is, at the tip of the "V" drawn by the boundary layer. Let $L1$ be the vertical distance between a horizontal line passing through the two points situated on the same contour line and the point of the valley bottom and $L2$ the horizontal distance between the point of the valley floor and a vertical plane containing the other two points and α the desired dip, we can then write:

$$\tan \alpha = L2 / L1$$

To illustrate this method, we will use the block diagram of an unconformity developed for Chapter 2 (Fig. 1.6). The horizontal plane containing the two points of the same altitude corresponds in fact to the unconformity plane. The low point is located at the bottom of the valley along the same layer boundary. The diagram gives the principle of the measurements

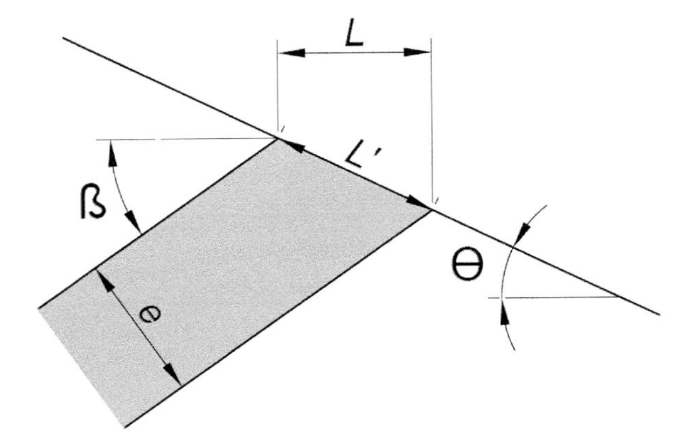

Figure 1.5 Diagram of the intersection relationships between a variable dip layer (β) and a variable slope (θ). This is how to find the thickness (e) knowing the width at the outcrop (L), the dip (β) and the slope (θ).

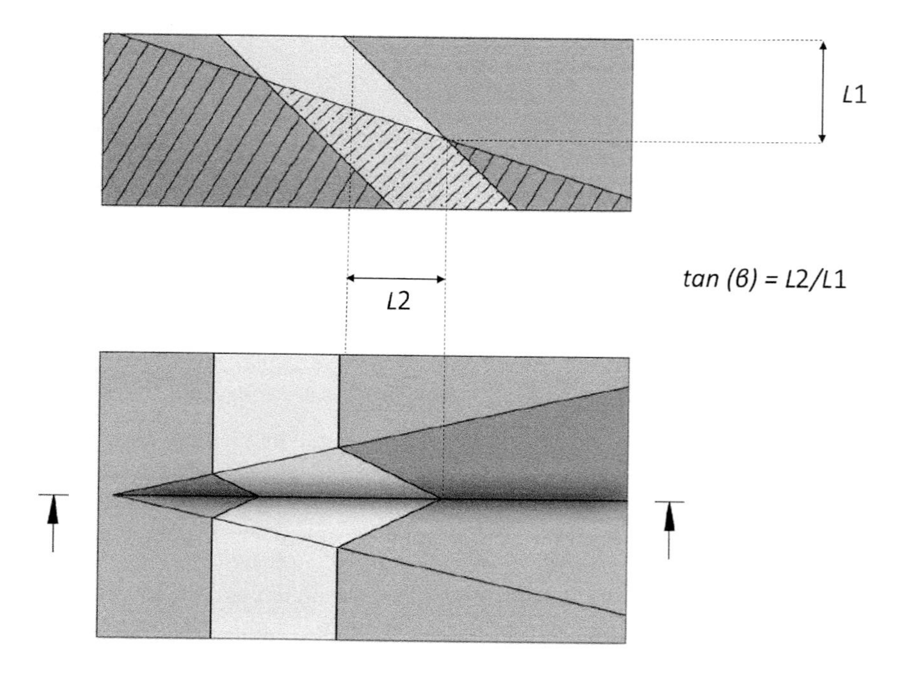

Figure 1.6 Map (bottom) and (top) combination of a section passing through the valley and a side view illustrating the three points method. The measurement of *L1* and *L2* allows calculating the value of the dip (α).

and calculation performed. On the map view (Fig. 1.6), we see the "V" shape of the layers in the valley. The tip of the "V" indicates the dip direction (but not its angular value). This finding is the "Rule of V's" as applied to valleys.

1.4 3D models of layers intersecting a topography

In nature, the dip of a geological layer or any other planar geological structure may vary from horizontal to vertical. The way this layer intersects the topography will tell us about its dip. We first propose a very simple model (Figs. 1.7, 1.8 and 1.9) representing a small circular hill with a single layer (representing a vein, for example) either horizontal, vertical or inclined. In all three cases, the representations are shown in map view (Fig. 1.7), in side view (Fig. 1.8) and in an oblique view (Fig. 1.9).

These theoretical models will have to be confronted with the reality either from photographs (Figs. 1.10, 1.11 and 1.12) or, and it is obviously better for this comparison, in the field.

The topography of the natural regions may consist of mounds and hills, but most often, it is thanks to the natural valleys that one can access the dip of the layers. For this, we designed a second model consisting of a stack of three layers intersected by valleys. By keeping valleys generally perpendicular to the layers (which is frequently the case), we can play on the shape of the valley (in "V" or "U" form, see Fig. 1.13) or on the direction of the dip of the layers (dip upstream or downstream, see Fig. 1.14).

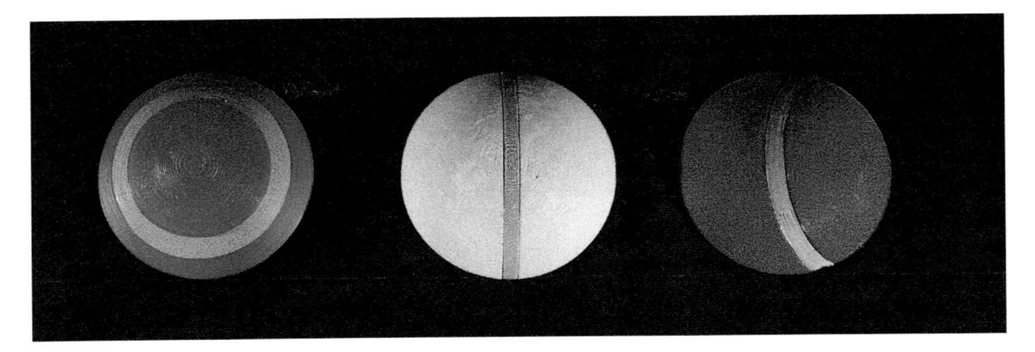

Figure 1.7 Photograph of printed models illustrating the intersection of a geological layer with a circular hill, seen from above (as on a map). From left to right: Horizontal layer, vertical layer and inclined layer.

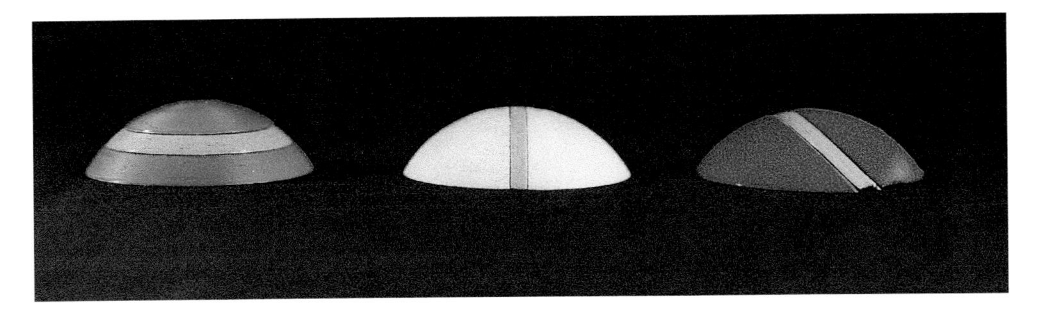

Figure 1.8 Photograph of printed models illustrating the intersection of a geological layer with a circular hill, side view. From left to right: Horizontal layer, vertical layer and inclined layer.

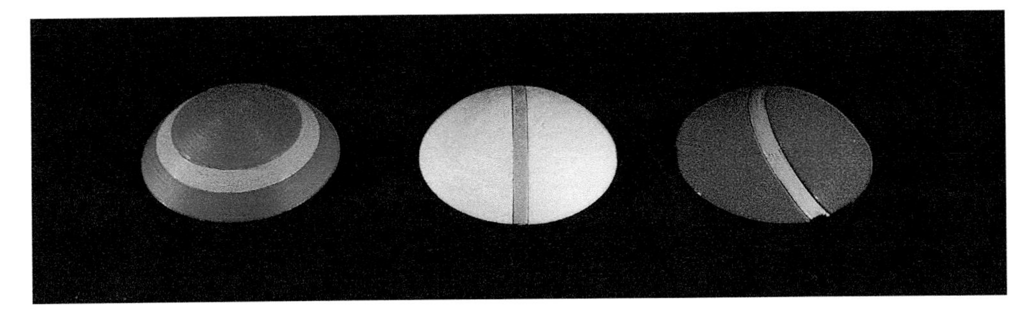

Figure 1.9 Photograph of printed models illustrating the intersection of a geological layer with a circular hill, oblique view. From left to right: Horizontal layer, vertical layer and inclined layer.

Figure 1.10 Model and field reality: Left-hand photograph, horizontal layers in the Kermanshah region (Iran). These are conglomerates and limestones of the Oligo-Miocene "Qom" formation lying flat on the structures of the Zagros range. Right-hand photograph, printed model illustrating the observed structure.

Source: (Left) © Dominique Frizon de Lamotte

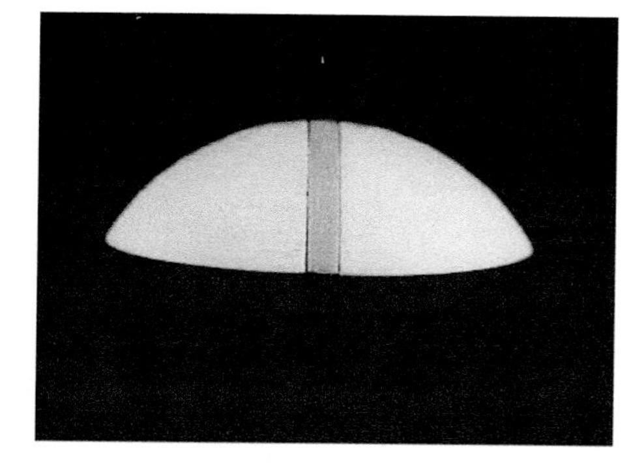

Figure 1.11 Model and field reality: Left-hand photograph, vertical layers (Cretaceous limestones and marls) in the Zagros range (Izeh region, Iran). Right-hand photograph, printed model illustrating the observed structure.

Source: (Left) © Dominique Frizon de Lamotte

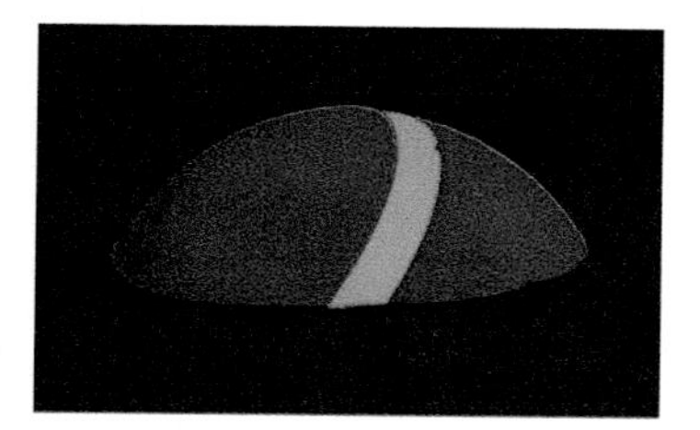

Figure 1.12 Model and field reality: Left-hand photograph, oblique layers (Miocene sands and silts) near the front of the Zagros range (Izeh region, Iran. Right-hand photograph, printed model illustrating the observed structure.

Source: (Left) © Dominique Frizon de Lamotte

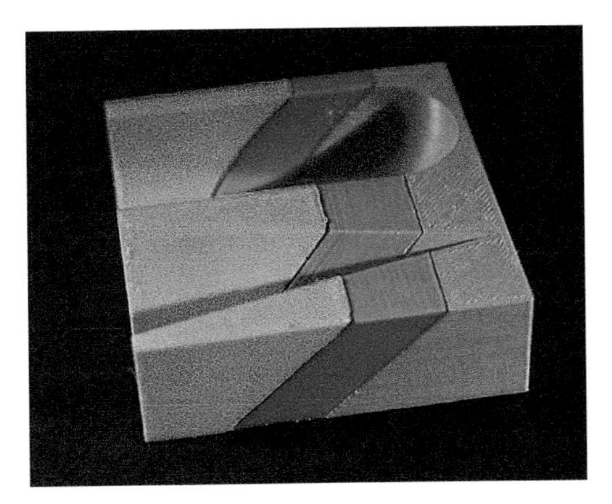
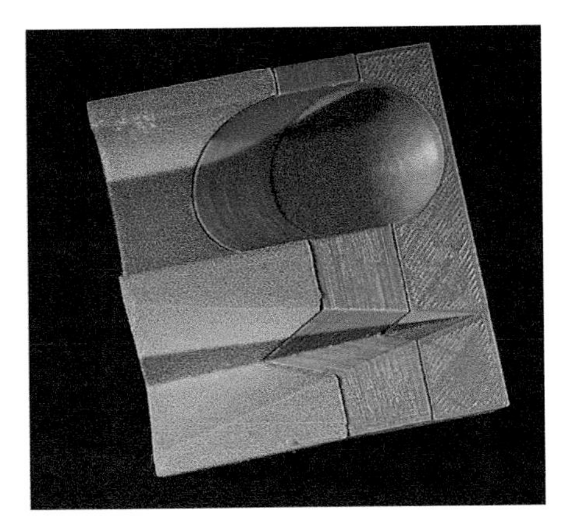

Figure 1.13 Influence of the shape of a valley (in "V" or "U" form) on the shape of the intersection between the topography and the parallel planes limiting the blue layer. On the left, the model is presented in oblique view of the printed block, and on the right in cartographic view.

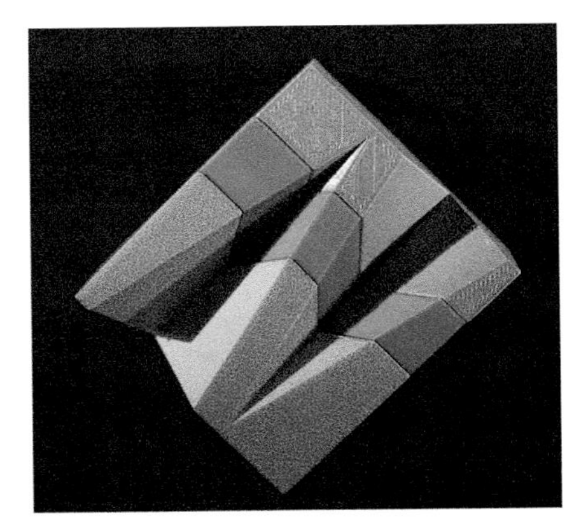

Figure 1.14 The printed model has a three-layer system intersecting two valleys arranged "head to tail." In the upper valley, the dip of the layers is downstream of the valley, and in the lower valley it is upstream. The left-hand photograph shows the model in oblique view, the one on the right in map view.

Valley morphology

Landforms are the result of interaction between erosive agents and vertical movements affecting the surface of the globe. Valleys are geographical depressions sculpted by rivers and glaciers during episodes of erosion (wind is also an eroding agent, but we will not deal with that in this text). In mountainous areas rivers carve V-shaped valleys while glaciers carve U-shaped valleys; let us see why.

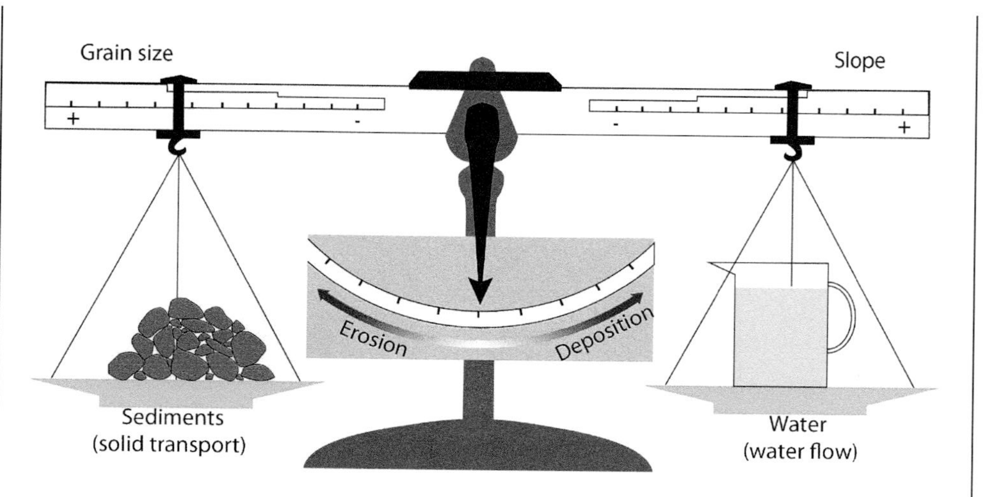

Figure 1A Lane's balance (1955) symbolizes the balance between the energy of the river and the work of the river. It shows how changes in solid load, transported particle size and the gradient or flow rate of the river determine whether the river cuts into its bedrock or deposits sediment.

From their source to their mouths, rivers adopt an "equilibrium profile" with a gradient that gradually decreases in a downstream direction. The equilibrium gradient at a given location corresponds to the gradient that allows water to flow (river energy) and sediments to be transported (work by the river). When this balance is disturbed (change in flow or solid load, change in gradient due to tectonic movements), either naturally or because of human activity, the river adjusts its bed by erosion or deposition of its solid load. The Lane diagram (Fig. 1A) illustrates this. If the work of the river is less than its energy, that is, if the flow is greater than the sediment load, the river uses the excess energy to cut into its bed and readjust its gradient. This cutting is vertical erosion that accentuates the difference in elevation between mountain ridges and valley bottoms. This vertical cut produces vertical slopes that are not very stable from a mechanical point of view. As a result, the slopes suffer a series of landslips that enable them to maintain an equilibrium gradient; the fallen material is then evacuated by the river and adds to its solid load. The combination of these two (vertical incision by the river and landslips on the slopes) gives rise to V-shaped valleys (Fig. 1B).

In mountainous areas glaciers flow under the influence of gravity from the heads of valleys, where snow accumulates and turns into ice, to valley bottoms that predate the glacier. The ice then forms large glacial tongues filling the valley bottoms. As ice accumulates, pieces of rock of all sizes are incorporated into the ice, by either falling from the slopes or being removed by the glacier. In contact with rocky surfaces, the particles that move with the glacier have a very powerful abrasive role (like sandpaper on wood) and sculpt both the bottom and sides of glacial valleys, inducing vertical deepening and widening of the valley, giving them a U-shape. The very steep walls of the valleys are maintained by the presence of the glacier, which exerts strong

Figure 1B The V-shaped Colca valley in Peru.

Figure 1C The U-shaped valley of Aosta in the Western Alps.

pressure until it melts. Most of the present valleys in the Alps, such as that shown in Figure 1C, are the result of this glacial erosion occurring during the Quaternary glaciations, when glaciers reached considerable thicknesses of several hundred meters. The last glaciation ended about 10,000 years ago and has left features that are a witness to its existence. But these are doomed to disappear gradually because the vertical flanks, which are no longer held up by the ice, are not mechanically stable and regularly collapse.

Chapter 2

When strata cross and overlie each other

The concept of unconformity

2.1 Definition and historical importance of the concept of unconformity

In geology, an unconformity is an ancient eroded surface truncating older tilted strata and supporting younger strata parallel to it. It could be said that the correct interpretation of an unconformity (Fig. 2.1) by James Hutton (1726–1797) announced the birth of modern geology. Indeed, Hutton understood that this particular geometric arrangement could only be obtained if: (1) the old strata were deposited horizontally, (2) they were then tilted and then

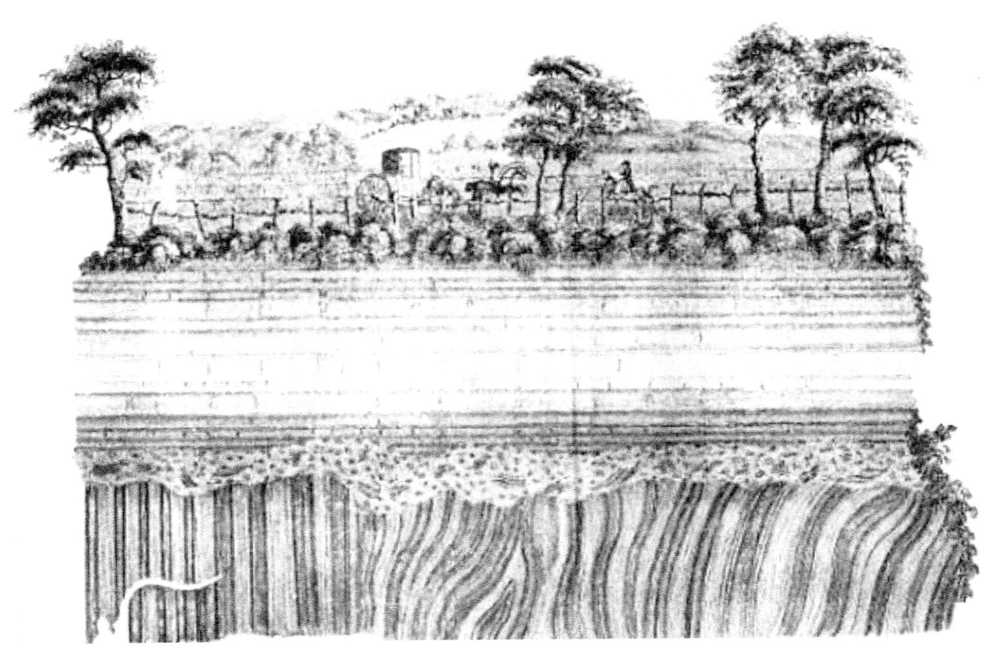

Figure 2.1 Illustration showing the unconformity at Jedburgh, Scotland, described by Hutton in his work "The Theory of the Earth." This is a regional unconformity known as the "Caledonian unconformity." Lower Paleozoic (Silurian) rocks are vertically tilted, then eroded and separated from the horizontal Upper Paleozoic (Devonian) rocks. Note the conglomerate just above the unconformity. This "basal conglomerate" is frequent but not always present at the top of the great regional unconformities.

(3) eroded and, finally, (4) a return of the sea led to the deposition of new horizontal strata. Why is this so important? Simply because Hutton's interpretation introduces the concept of a cycle (in this case, a sedimentary cycle) into geology. This broke with the idea that the sedimentary process was inevitable, resulting from a gradual settling of sediment at the bottom of a primordial ocean, as proposed by the Neptunist school (after Neptune, the Roman god of the sea) of the followers of Abraham Werner (1750–1817). In addition, this interpretation showed that the process can be interrupted by deformations leading to uplifts and exposure to open air and consequently erosion. The resulting erosion can then feed sedimentation into an adjacent basin and so on. Hutton also observed that *plutonic rocks* could cut out sedimentary rocks and therefore be younger. In general, he considered that the "central fire" played a major role, hence the name "plutonism" (after Pluto, Roman god of the underworld) given to the school of thought that he belonged to.

The cyclical nature of geological processes provides another way of conceiving geological time: no longer short and polarized, as the "Neptunists" thought, but rather cyclical and infinitely long: "No vestige of a beginning, no prospect of an end" (Hutton, 1795). We now know that the earth's timescale, even though very long (4.53 Ga), is not infinitely long. There was a beginning and there will be an end. And neither is time monotonous, a kind of eternal cycle; we now know that it is punctuated by catastrophes.

Geological time and philosophy

In what way could the use of the concept of time in geology have a philosophical significance? Philosophy is driven by a will to know, which it shares with science, and sets itself the general task of questioning the problematic aspects of human experience. Now if there is indeed one aspect of our existence that raises questions, it is time, and more particularly the *measurement* of time, because measuring means establishing a quantifiable relationship between two comparable magnitudes: the aim of the measurement is to relate what is being measured to an appropriate unit. But what unit is to be chosen for time? How can something that, by definition, is elusive be determined? Societies that have measured time have relied (and still do) on the repetition of natural, supposedly regular phenomena. Thus, the days and the alternation of the seasons are events corresponding to segments of our existence, based on our activity and serving as a basis for establishing units of duration. The measurement of time thus primarily serves an *anthropological* interest: it comes from the needs of human labor and from the relationship that mankind has with its environment. To elaborate: all societies, without exception, feel the need to think in terms of time periods that go far beyond the experience of a single human life. This is the case as soon as they form a narrative of a past that goes back beyond three or four generations. They then have a paradoxical dimension, which often goes unnoticed since it is so ingrained in our ways of thinking: the *archaic*. Anthropology terms as "archaic" a past that is more distant than any lived past, which cannot therefore be recalled by memory. This dimension is what gives human societies the mythological accounts of their origins through which their existence can be justified. The archaic, as indicated by its Greek origin "arche" (which means "beginning" or "first principle") is undoubtedly an imaginary past, but also a founding past.[1]

It is surmised that geology modifies these traditional ways of conceiving time. Like other natural sciences, it treats time as a dimension in which events unfold. Time is thus represented as a *line* along which the evolution of the phenomena being witnessed – or which are reconstructed from their traces – can be followed. This representation, so familiar that it now seems quite innocuous, nevertheless carries with it a powerful ability to *demystify*. Indeed, following Hutton's famous phrase ("no vestige of a beginning, no prospect of an end"), the indefinite line of time plunges into the past without ever finding its point of origin or beginning – except in the farthest reaches of the creation of the universe. It thus precludes mythological fabrications and makes this archaic dimension of the past literally disappear, despite it seeming indispensable to all societies. Taking this further: geology does of course use traditional time units, but it also introduces us to timescales that are so immense that they are incommensurable with our existence, and no lived experience can demonstrate their extent. It produces a sort of *decentering of the measurement of time*, which is no longer proportionate to human activities but derives from gargantuan natural processes that are indifferent to mankind.[2] The use of the concept of time in geology thus imposes a double loss: the loss of the reference to a founding past and the loss of a temporality dependent on anthropological needs.

This double abandonment is not without recompense. By conceiving time as an indefinite line, the geologist is allowed to question the past with the certainty that it is *intelligible*. He presupposes that the events that fill the line (even if they are only partially known) are linked in a determined and explicable order. This representation of time allows the geologist to reconstruct the past from signs that are deposited in the present, to arrange them into an order and thus to establish an authentic historical framework. In short, the use of time in geology is the condition that makes a rational inquiry into the past possible, a *true account* of the events that have affected and still affect the earth.

Julien Douçot Professor of Philosophy
at Paul-Eluard Secondary School, Saint Denis,
France. Bachelor student of Earth Sciences at
CY Cergy Paris Université

1 A term with the same root is used in geology: the "Archaean" is the earliest era in the history of the earth of which we have traces.
2 The decentering accomplished by the natural sciences has been a philosophical issue since the 17th century, especially with Pascal. A summary is to be found in a renowned text by Freud on the "three wounds" inflicted on human narcissism, cf. *A General Introduction to Psychoanalysis*, II, 18, Paperback edition.

2.2 Overview of unconformities, the concept of basement and cover, great regional unconformities

2.2.1 *Generalities about unconformity*

Unconformity is the opposite of conformity. Conforming strata are parallel to each other, and a simple hiatus (gap in deposition or erosion) is not enough to define an unconformity. It also requires a truncation (overlap) of the underlying strata; otherwise, it is no more than a discontinuity. To emphasize this aspect, it is sometimes specified as an angular unconformity. It must also be pointed out that an unconformity is a stratigraphic contact and not an "abnormal contact," that is to say a tectonic contact, otherwise known as a fault, as is sometimes mistakenly believed.

2.2.2 *Basement and cover*

Unconformities observed locally on outcrops (Fig. 2.1) may have a wide-reaching significance. Indeed, the study of continental domains highlights the existence of great regional unconformities structuring landscapes. They separate ancient truncated mountain ranges from large sedimentary basins that have undergone little tectonic deformation. These form the cover (sedimentary cover) as opposed to the basement located underneath (Fig. 2.2). The basement is generally poly-deformed, metamorphosed and "intruded" by granites. Its mechanical behavior is therefore different from that of the cover. This has important consequences when the basement and the cover are engaged together in an orogenic process. The cover may fold, whereas the rigid basement is more likely to develop faults. This diversity of behavior leads to the development of a *décollement* at, or close to, the basement-cover interface when the region is involved in large tectonic deformations (see Chapters 4 and 5).

2.2.3 *Great regional unconformities*

Great regional unconformities generally bear the names of the truncated orogenic systems, which they overlie. Thus, in Scandinavia and Scotland, the "Caledonian unconformity" discovered by Hutton separates the Caledonian chain, formed in the Lower Paleozoic, from the Upper Paleozoic sedimentary rocks preserved in intracontinental sedimentary basins. In Western Europe, and France in particular, the "Hercynian unconformity" separates the Hercynian (or Variscan) chain, formed in the Upper Paleozoic, from large Mesozoic basins: the Paris basin, the Aquitaine basin whose southern edge is integrated in the Pyrenees mountain range, or the southeast basin now integrated almost entirely into the Provençal and Alpine mountain ranges (Fig. 2.3).

A west-east cross-section through the Paris basin (Fig. 2.4) illustrates the dual vision of geology (map and cross-section). The cross-section shows the "Hercynian unconformity"

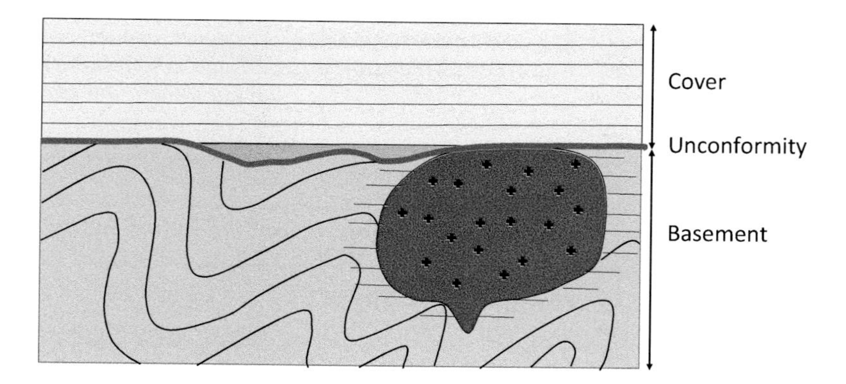

Figure 2.2 Theoretical cross-section illustrating the concepts of basement and cover. The sedimentary cover (shown in yellow) is separated from the basement by an unconformity (blue line). The unconformity surface is an old surface of erosion, hence its irregular aspect. Pockets of conglomerate (dark yellow) are often preserved at the base of the cover. The basement consists of folded and faulted sedimentary rocks, often metamorphosed and intruded by magmatic rocks such as granites (in red). The intrusion of granites results in the formation of a contact metamorphic aureole (horizontal dashes).

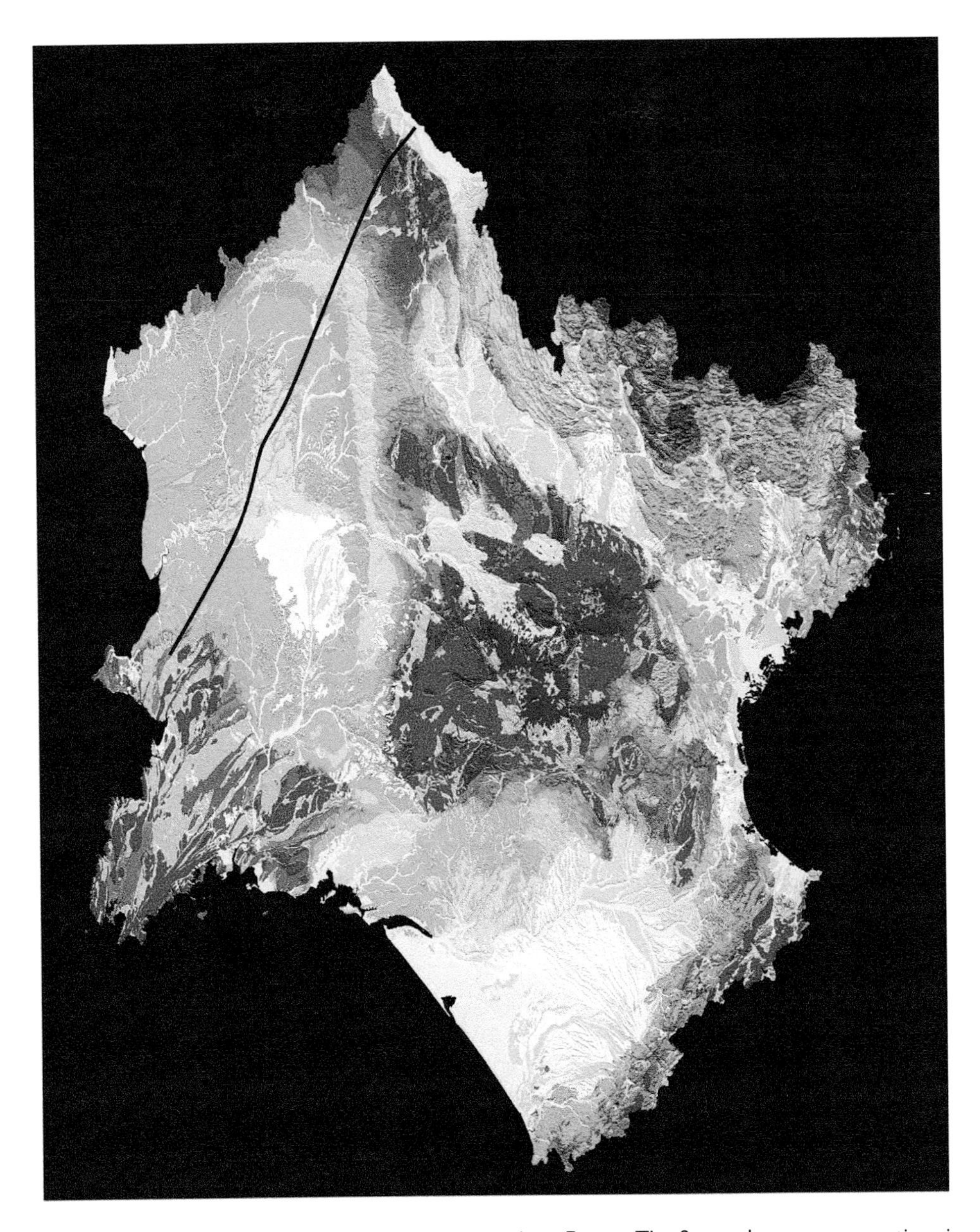

Figure 2.3 The "Hercynian unconformity" in metropolitan France. The figure shows a perspective view of the geological map of France overlying a digital terrain model. The Mesozoic terrains in purple (Triassic), blue (Jurassic) and green (Cretaceous) lie unconformably on the Variscan basement consisting of deformed Precambrian and Paleozoic sedimentary rocks, granites (in red) and metamorphic rocks. The "Hercynian unconformity" borders the Armorican massif to the east. It also surrounds the Massif Central and the Vosges. This unconformity is a major structuring element separating the sedimentary cover from its basement. In the recent chains (Pyrenees, Alps), the Variscan basement is to be found at the heart of the chains but reworked by Cenozoic (Alpine) deformation.

Source: © BRGM, France, document downloadable from the website: www.brgm.fr/editions/ouvrages-cartes-brgm-editions

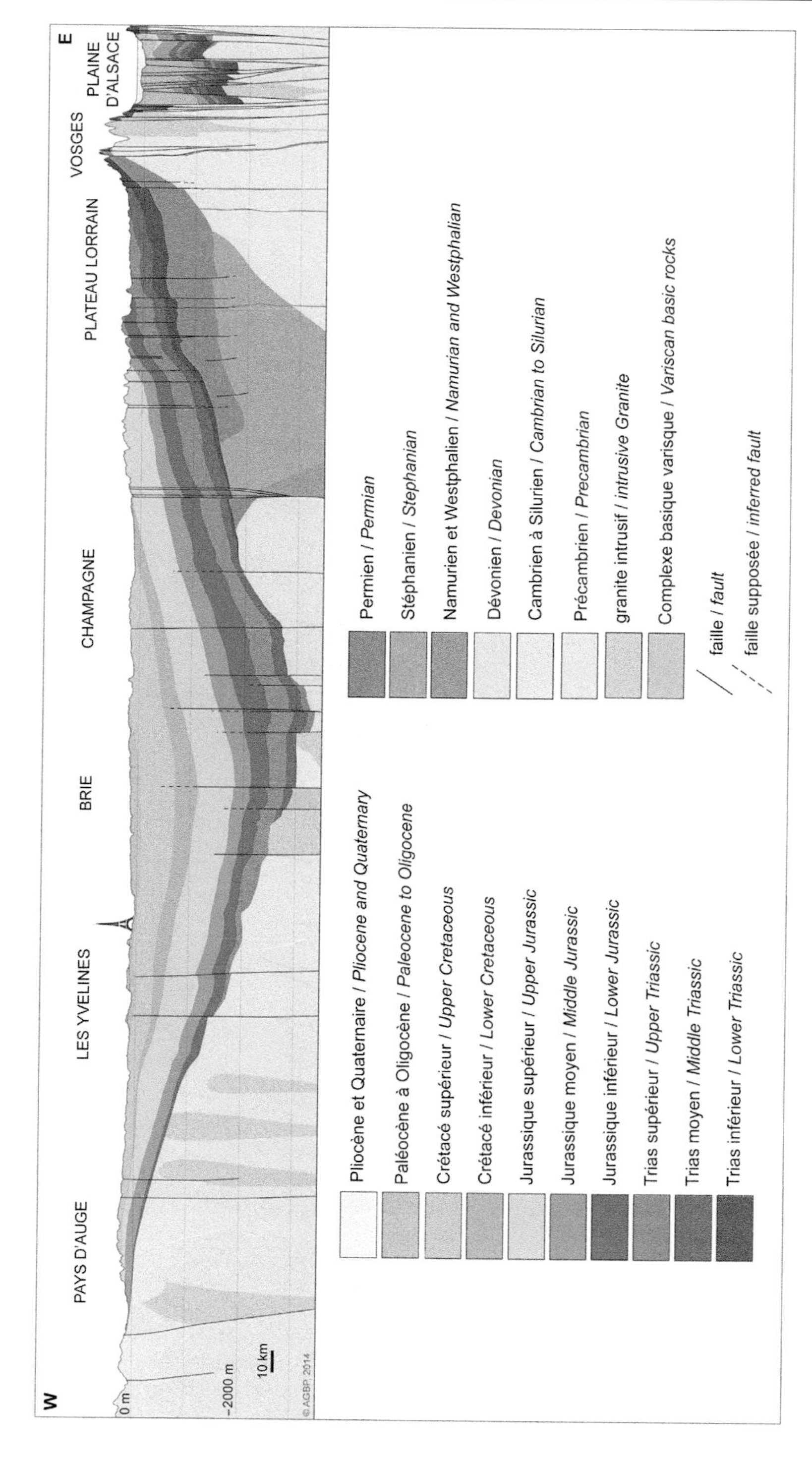

Figure 2.4 West-east cross-section of the Paris basin drawn up by the Paris Basin Geologists Association ("Association des Géologues du Bassin de Paris," APBG). Besides showing the Hercynian unconformity and the organization of the Mesozoic sedimentary terrains, the cross-section shows a thick Stephanian-Permian basin sealed by the Lower Triassic to the east.

Source: Cross-section available online (www.agbp.fr/blog/2017/10/recevez-gratuitement-la-coupe-du-bassin-parisien-en-a4#.WjjjBFXiapo)

at the base of the Mesozoic sediments. Underneath this, the Precambrian (Icartian) and Paleozoic sediments are folded together and interspersed with magmatic rocks. Note that the cross-sectional view (Fig. 2.4) is geometrically similar to the map view (a basic exercise for students in earth sciences is building a cross-section from the interpretation of a geological map). To the west of the Paris basin, Mesozoic sediments truncate the folded terrain of the Armorican massif (Fig. 2.3) in the same manner to that shown in the cross-section.

Here there is the need to stress another essential point provided by both cross-section and map. The cross-section (Fig. 2.4) shows the erosional truncation of the basement strata but also an apparent truncation of the sedimentary layers at the top of the unconformity. This indicates that the strata directly overlying the unconformity are younger in the west than in the east, as if the basin progressively overflowed westwards over time. The map view (Fig. 2.3) even indicates that the Upper Cretaceous layer (light green) may lie directly on the basement of the Armorican massif. This configuration is indicative of a transgression of the Mesozoic sea from east to west, culminating in the Late Cretaceous by the deposition of chalk. There was an ocean to the east (the Tethys) that has since disappeared, while there was still no Northern Atlantic to the west.

In detail, the cross-section of the Paris basin also shows a very thick basin of Stephanian (Upper Carboniferous) and Permian deposits on the east side (Vosges). Although not clear in the cross-section, this basin is a "rift" basin that is a basin created by extensional deformation and limited by normal faults (see Chapter 3). The Hercynian unconformity as defined here is therefore a composite discontinuity combining a post-orogenic history (with respect to the Variscan chain) and a post-rift history (with respect to the Stephanian-Permian rifting). As orogenic processes are followed by the extensional collapse of the newly formed mountain range, this composite (polygenic) character of the great regional unconformities is very general.

In order to understand the general nature and the importance of the concepts of basement and cover, the whole European continent needs to be examined. In Europe, as in the other continents, the age of the basement differs according to the place, and so, consequently, does the age of the cover which it supports (Fig. 2.5). To the east, the basement constituting the "Scandinavian shield" is Precambrian. It supports sedimentary rocks ranging in age from Neo-Proterozoic to Cenozoic, constituting the East-European or Russian platform (in this case, the term platform is used as a synonym for cover; these are in actual fact vast plateaus). To the northwest, that is, primarily in Norway and Scotland, the basement is made up of the Caledonian chain, formed as seen earlier at the end of the Lower Paleozoic. The cover, which must of course be younger, developed by the Devonian. In southern Europe (Spain) and western-central Europe (France, Germany), the basement is Variscan, that is, Late Carboniferous. The cover is essentially Mesozoic and Cenozoic (Fig. 2.3 and 2.4) although, as seen previously, the first post-Variscan deposits date from the Permian and even from the Upper Carboniferous at the center of the Variscan chain.

These age differences in basements were understood quite early on in the history of geology. Since the planation of chains occurs in a continental context and the first deposits were sands and red sandstone, Europe previously saw the formation of the Old Red Sandstone and New Red Sandstone continents. The old red sandstones are Devonian in age and were deposited on the top of the Caledonian chain after its planation. The new red sandstones are Permian to Triassic in age and developed on the top of the Variscan chain after its planation. The former are the stones found in old Glasgow, while the latter are those of old Strasbourg.

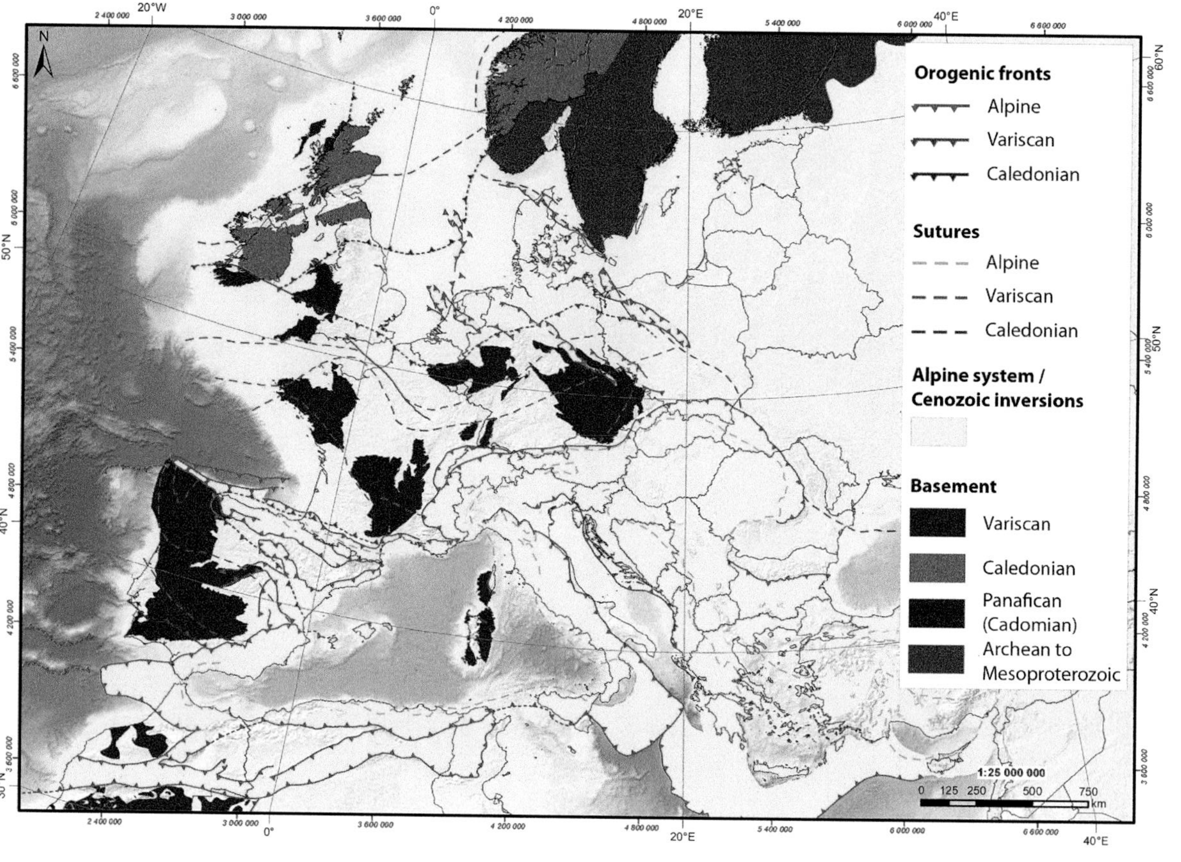

Figure 2.5 Highly simplified structural map of the European continent showing the basement and the sedimentary cover (in different colors and gray, respectively). The age of the basement varies: Archaean to Mesoproterozoic in the Baltic Shield; Caledonian (Lower Paleozoic) in Norway, Scotland and Ireland; Variscan in Germany, England, France and Spain; and Pan-African (Upper Proterozoic) in the blocks caught between the Variscan chain and the Caledonian chain (the so-called "Avalonia terrane" cropping out in England and Belgium). The cover lying directly on the basement dates from the Late Precambrian, Devonian and Permian, respectively. Also shown: the Cenozoic chains (Alps, Pyrenees, Apennines, Carpathians, etc.) and narrow zones deformed at that time (North Sea, Poland) in yellow. On the map, the basins are represented uniformly by the same gray. However, the age of the base of this cover depends on the age of the underlying basement. It cannot be older but may, in some cases, be much younger (see Fig. 2.3).

2.3 Great ocean unconformities, the concept of a "break-up unconformity"

For reasons related to advances in exploration techniques, the geometry of the continental margins of the oceans was not fully understood until the second half of the 20th century, following on from the results of the search for hydrocarbons by oil companies and of deep drilling for academic purposes (IPOD, then IODP programs). These subsurface data reveal that the margins known as rifted or passive margins, that is, currently tectonically inactive, are a result of the thinning of the continental lithosphere, leading to its break-up and the creation of an oceanic space. In the upper crust, thinning is accommodated by the development of normal faults that remain active until the crust – and even the continental lithosphere – breaks. During this first phase, which corresponds to the rifting process, subsidence (the gradual sinking of the substratum making a space available for sedimentation) is controlled by the action of faults. As soon as the break-up occurs, the gradual cooling of the margin and the adjacent oceanic domain leads to a second subsidence phase, known as thermal subsidence, during which the creation of space available for sedimentation is no longer controlled by faults but by the progressive cooling of the lithosphere. The transition between these two regimes is underlined by a great discontinuity known as a "break-up unconformity" (Fig. 2.6).

In actual fact, this break-up unconformity is nothing more than a "post-rift" unconformity (Fig. 2.6) similar to that found over a rift system, even if this one failed, that is, it did not go as far as to rupture the crust. A post-rift and, by extension, a break-up unconformity does not necessarily correspond to a generalized erosional surface. Indeed, in some cases a complete uplift only occurs at the end of the rifting process. This is the case of the Permian rifting preceding the establishment of the Paris basin (Fig. 2.4) but it is not systematic. In all cases, however, it is a major discontinuity truncating the normal faults developed during the rifting episode. It should be noted that some faults may continue to move after rifting. This is nonetheless moderate movement on a completely different scale than the movements observed during the rifting period.

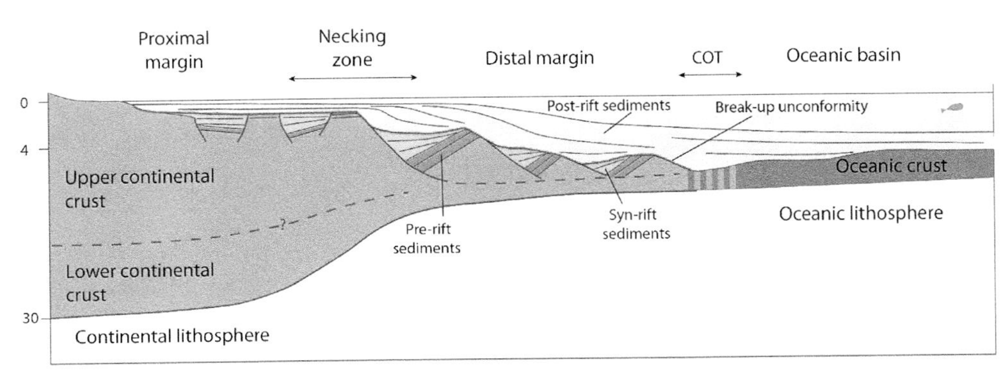

Figure 2.6 Theoretical cross-section of a rifted (passive) margin showing the break-up unconformity (red line) sealing the old normal faults that led to the thinning of the continental crust before its rupture. It should be noted that the post-rift unconformity of the proximal part of the margin is often older than that of the distal margin, which actually corresponds to the break-up unconformity.

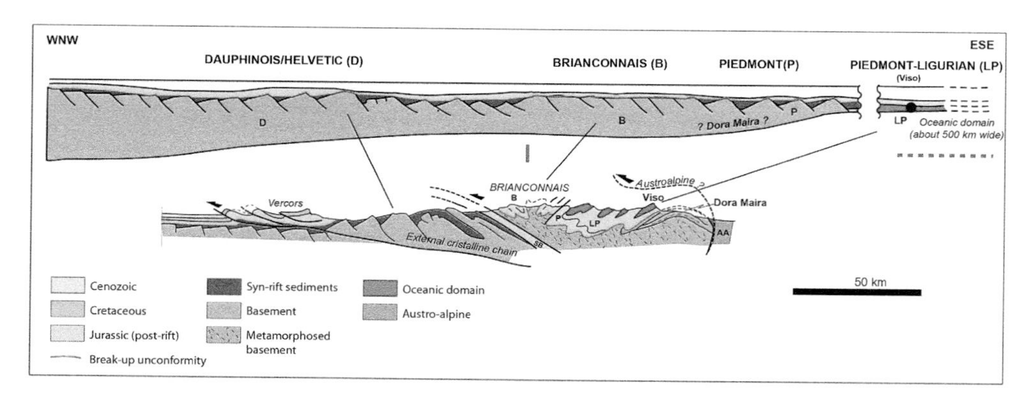

Figure 2.7 Schematic structure of the Alps (bottom) and its geometry restored at the end of the Jurassic (top). The blue layers of the Upper Jurassic correspond to the first "post-rift" sediments, covering the break-up unconformity and constituting the first sedimentary cover common to the distal margin and the oceanic domain. At the front of the internal zones, the acronym SB corresponds to the sub-Briançonnais paleogeographic domain.

Source: Modified according to Ph. Agard and M. Lemoine, Visages des Alpes, CCGM/CGMW, Paris

Collision chains are formed at the expense of an oceanic domain that has disappeared and has been sutured, while intracontinental chains are formed at the expense of a failed rift. Traces of the break-up or post-rift unconformity marking the end of extensional tectonics are to be found in both configurations. In the Alps, the prototype of a collision chain, this unconformity was recognized in the 1980s and corresponds to the base of the cover that is common to both the old oceanic domain and the old passive margin (Fig. 2.7).

When the conditions are favorable, inherited normal faults can be seen under this unconformity. These faults were sealed and preserved during the tectonic inversion (that is, the transition from extension to compression) that eventually led to the formation of the mountain range. Thus, two ancient unconformities can be identified in the Alps: the "Hercynian" unconformity on top of the Variscan basement and the break-up unconformity within the Jurassic strata. There are a number of local unconformities testifying to the compressive episodes. On the other hand, there is no "alpine" unconformity as yet. This will only be seen once the chain has been completely eroded and truncated.

2.4 Dating an unconformity

We have seen that great regional unconformities in the continental domain were most often composite, that is to say, they accumulated successive events on the same surface. In theory, it is quite simple: the age of an unconformity is determined by a time interval limited by the age of the oldest layer on the unconformity and the age of the most recent rock below the unconformity. In practice, it is more complicated and often requires correlations over long distances. This is well illustrated by the cross-section of the Paris basin (Fig. 2.4). It is clear that the time period of the western part of the cross-section (between the Lower Paleozoic and the Jurassic, that is, at least 200 million years) is much greater than the period to the east (between the Permian and the Triassic). In this instance, it is on the order of a million years because there is in effect a sedimentary hiatus between the Permian and the Triassic, even though these two periods are contiguous.

This point of view must be qualified since, as noted earlier, there are two unconformities to the west that are merged into a single discontinuity, whereas these two unconformities are separated by a Stephanian-Permian basin to the east. The great abrasion surface, most likely fairly flat, on which the Mesozoic rocks were deposited was formed subsequently, after not only the building but also the collapse of the Variscan chain.

Historically, the concept of unconformity has been used to provide a chronological framework for the major orogenic cycles. This is easily explained because they are the most obvious in the continental domains, which were the first to be explored. In addition, they determine a number of elements of physical and human geography. Conversely, "post-rift" unconformities are more often concealed under large sedimentary basins and cannot therefore be observed without subsurface data (seismic, wells), except when old rifts and continental margins are integrated into mountain ranges (Fig. 2.7).

This can be seen in Africa, more precisely its northern part, taken together with Arabia, which shares a long common history with Africa (Fig. 2.8). There are a number of different overlapping basins. Figure 2.8 shows these basins categorized by age. A Proterozoic basin can be seen in the west, the Taoudeni basin that covers the West African craton and has Paleozoic sediments at its summit and even a thin layer of Mesozoic sediments. Elsewhere, the basement was formed during the Pan-African orogeny (late Precambrian) and the cover is Phanerozoic (Paleozoic and later). By removing the Meso-Cenozoic cover, a "subcrop map" can be constructed, which reveals an "arch-and-basin" architecture, very typical of the old continents (Fig. 2.9).

This structure is sealed by a major unconformity, which cannot be fully characterized without the aid of subsurface data. Thus, the structure needs to be uncovered gradually, starting from the top. The Cenozoic at the top of the pile can be ignored. In Arabia, the Mesozoic is complete and concordant on the Permian and even on the Upper Middle Carboniferous (Fig. 2.10). The latter, on the other hand, lies clearly unconformably on the older terrains, even though this unconformity is not very pronounced (the angle between the layers underneath the unconformity and the unconformity itself is weak or nonexistent, such that the unconformity can only be seen by "de-zooming" considerably).

The more pronounced unconformity within the cover is therefore pre-Carboniferous in these regions (Fig. 2.10). It has often been correlated with Europe's "Hercynian unconformity," suggesting that the arches-and-basins architecture was generated under a compressional regime. However, this is a long way from the Variscan front along the Atlantic at the western tip of Africa (Fig. 2.8 and 2.9). Furthermore, recent data show that this pre-Carboniferous unconformity seals some extensive structures developed by jerks during the Paleozoic (Fig. 2.11), with a major event in the Cambrian and another in the Upper Devonian.

This unconformity is present throughout North Africa and Africa. In the western regions, toward Morocco and western Algeria, its significance is obscured because other events, also generating unconformities, complicate the reading of the signal. First there is the formation of the Variscan chain during the Upper Carboniferous, then the various rifting phases giving birth to the Central Atlantic, the sedimentary basins that currently form the Atlas mountains and the Maghreb branch of the Tethys, the ocean that formerly separated Africa from Eurasia.

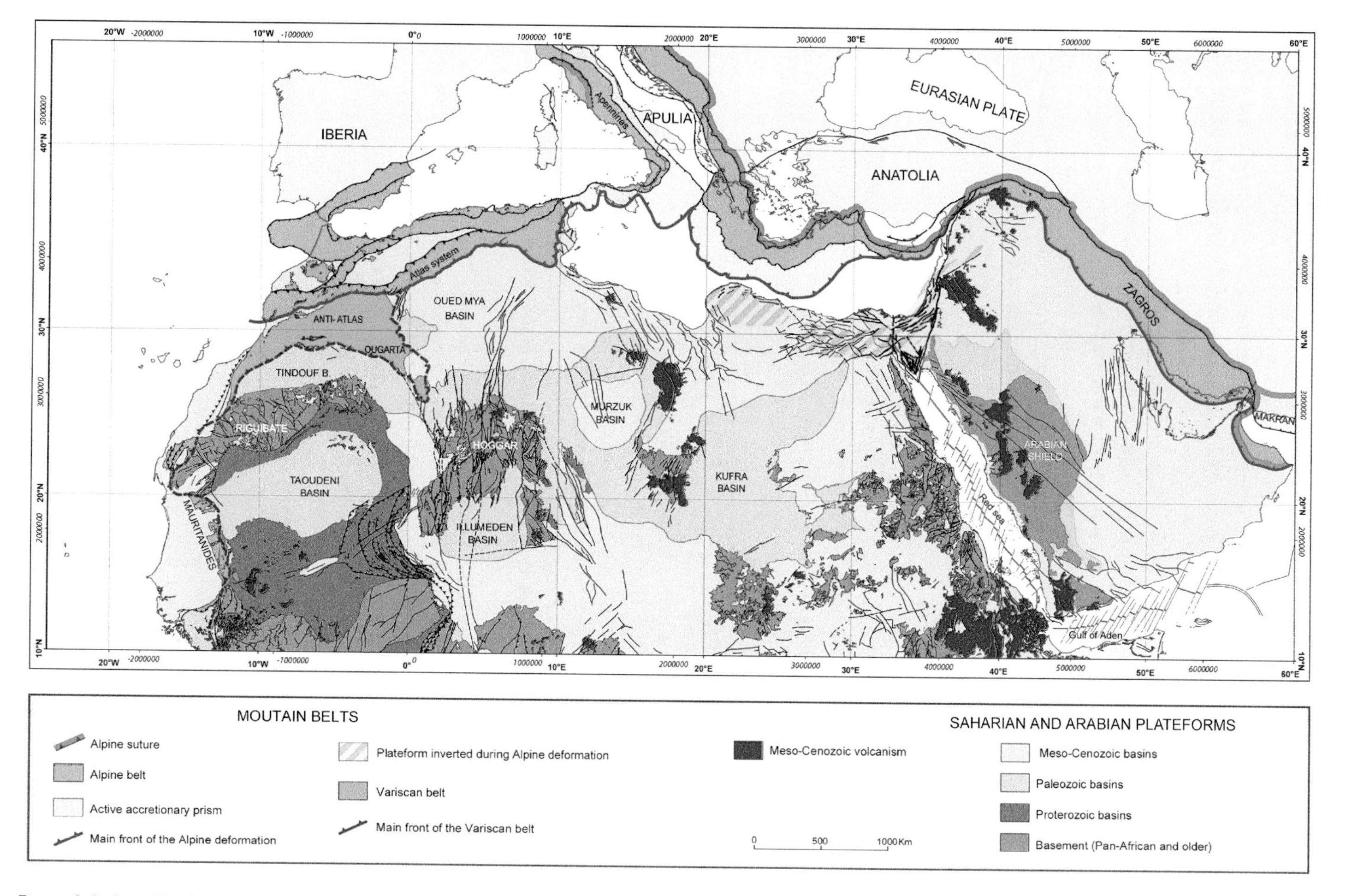

Figure 2.8 Simplified structural map of northern Africa and Arabia. To the south of the "alpine front" (blue line) and east of the Variscan front (red line), there is an undifferentiated basement (in gray) and Proterozoic, Paleozoic and Meso-Cenozoic covers, respectively.

Source: From Frizon de Lamotte et al. (2013), modified

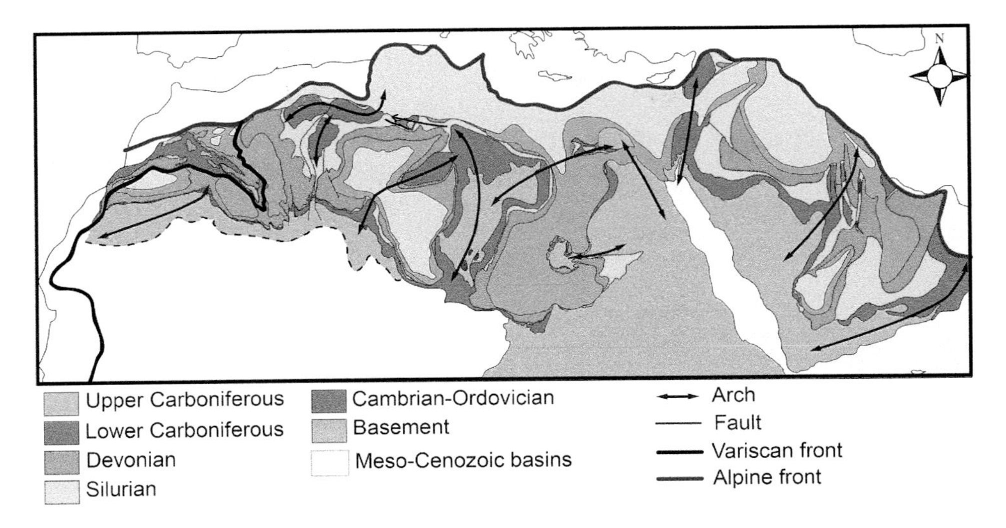

Figure 2.9 Sub-crop map under the Mesozoic (map created by removing the Mesozoic sediments) revealing the arch-and-basin architecture characterizing the interior of the continent.

Source: From Juliette Rat, unpublished Master 2 Thesis

Figure 2.10 Photo from a helicopter showing the pre-Carboniferous unconformity in the High Zagros range (Iran). The Lower Paleozoic terrains on the valley floor are unconformably covered by a stack of limestone rocks that run continuously from the Permian to the Upper Cretaceous. In the foreground, Upper Cretaceous limestones are overthrusted (thrust fault in the river) by Lower Paleozoic rocks and their cover.

Source: Photo © Dominique Frizon de Lamotte

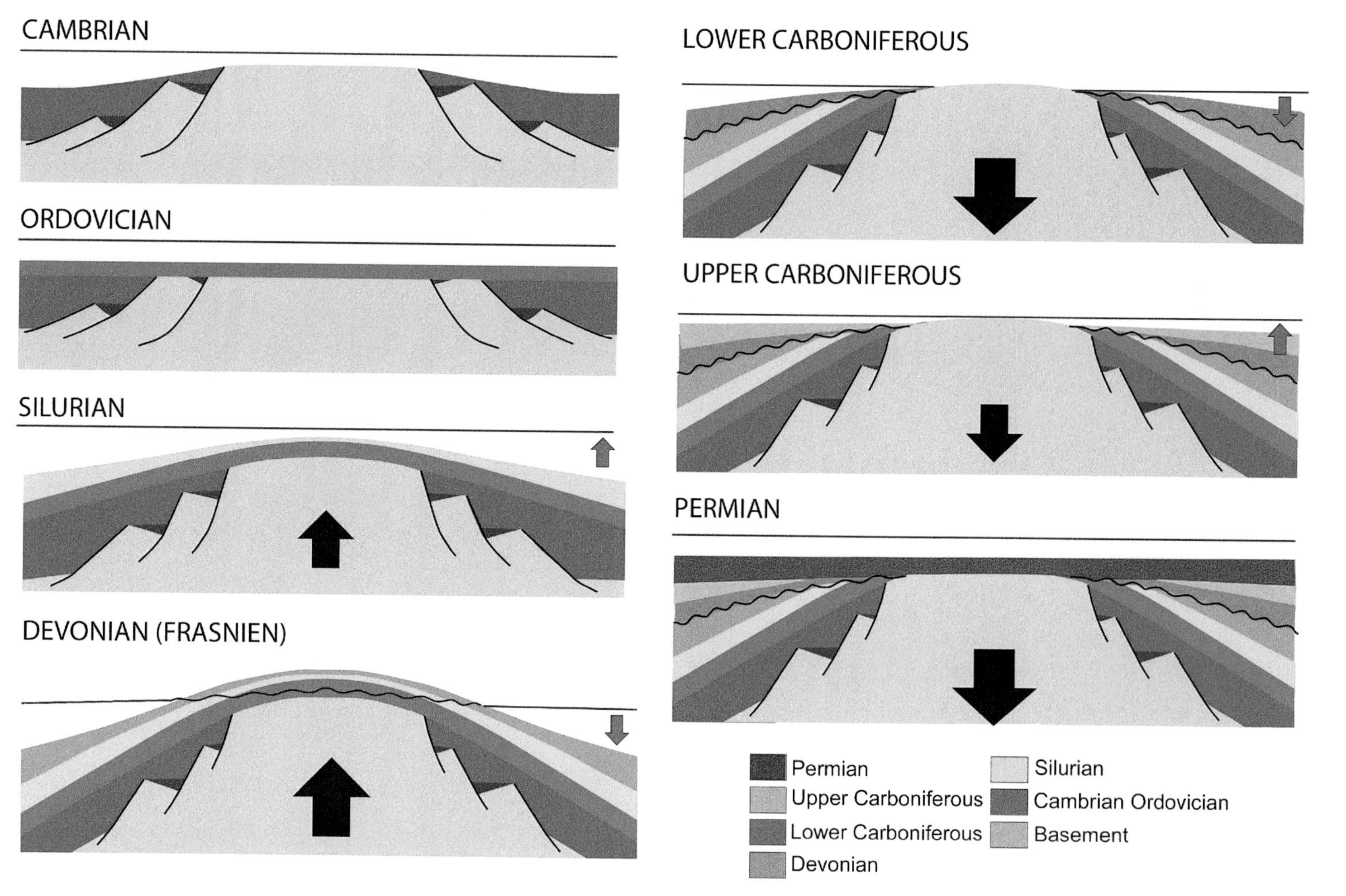

Figure 2.11 Conceptual model illustrating the formation of the arches of the Saharan and Arabian domain. The black arrow indicates the movement of the basement (uplift or subsidence); the blue arrow indicates eustatic movement (rise or fall of the global sea level).

Source: From a drawing by Juliette Rat, unpublished Master 2 work

In the oceanic domain, the absence of a significant hiatus makes it easy to date a break-up unconformity (provided it can be accessed). There is, however, a conceptual difficulty: can an ocean be envisaged as opening all of a sudden at exactly the same time over its entire length? This seems to be more or less the case for the large linear oceanic domains such as the Central Atlantic, for instance, but it becomes difficult to conceive when the opening takes place in a triangle whose tip spreads progressively (propagator concept). It was pointed out earlier (Fig. 2.6) that the age of the post-rift unconformity may be older on the proximal than on the distal margin. In this case, the explanation is simple; after a stage of diffuse deformation, the later phases of the rifting are localized close to the place where the continental break-up will occur.

2.5 Dating the duration of an event

In general, an unconformity signifies the end of a process (end of an orogenesis, end of a rifting period) but only gives a rough idea of its duration. This duration can be measured in some cases and under specific conditions. The process must be recorded continuously by sedimentary records, which must be dated. It requires continuous subsidence plus a marine environment that is rich in good stratigraphic fossils to allow precise dating. Figure 2.12 shows two examples: a normal listric fault (spoon-shaped) and an anticline fold. In both cases the development of growth strata forming “progressive unconformities” can be seen. They indicate a contemporaneity between deposition and the creation of the structure. These growth structures are common in rift basins (in extension) and in flexural basins (in front of mountain ranges). In both situations, subsidence is generated by the geodynamic context. It may be the thinning of the continental crust (rifting) or the bending of the lithosphere due to the excess weight caused by crustal thickening.

The southern flank of the Pyrenees in Spain is particularly rich in progressive unconformities. Figure 2.13 shows an example in front of a famous anticline: the Sant Corneli anticline.

2.6 Unconformity model for 3D printing

The first model proposed is inspired by the impressive unconformity exposed in the immediate vicinity of the village of Minerve (Hérault, France). The upper strata forming the cliff are carbonate Eocene in age. They overlie an unconformity that truncates the Paleozoic strata outcropping into the valley. These toppled over before being eroded (Fig. 2.14).

The proposed model comprises five parts that can be nested in a “dovetail” system (trapezoidal pin fitting into a groove of the same shape to provide a sliding connection). The

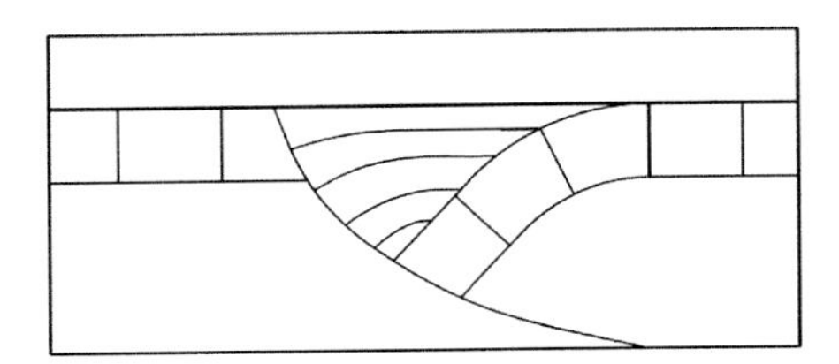

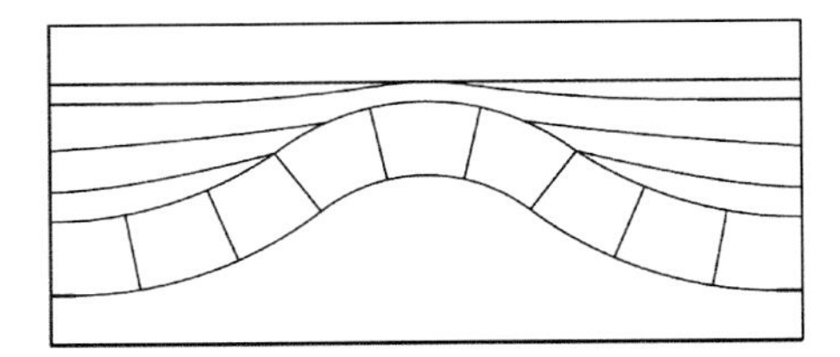

Figure 2.12 Examples of progressive unconformities in an extensional setting on top of a listric fault (left) and in a compressional setting on top of an isopach fold (right).

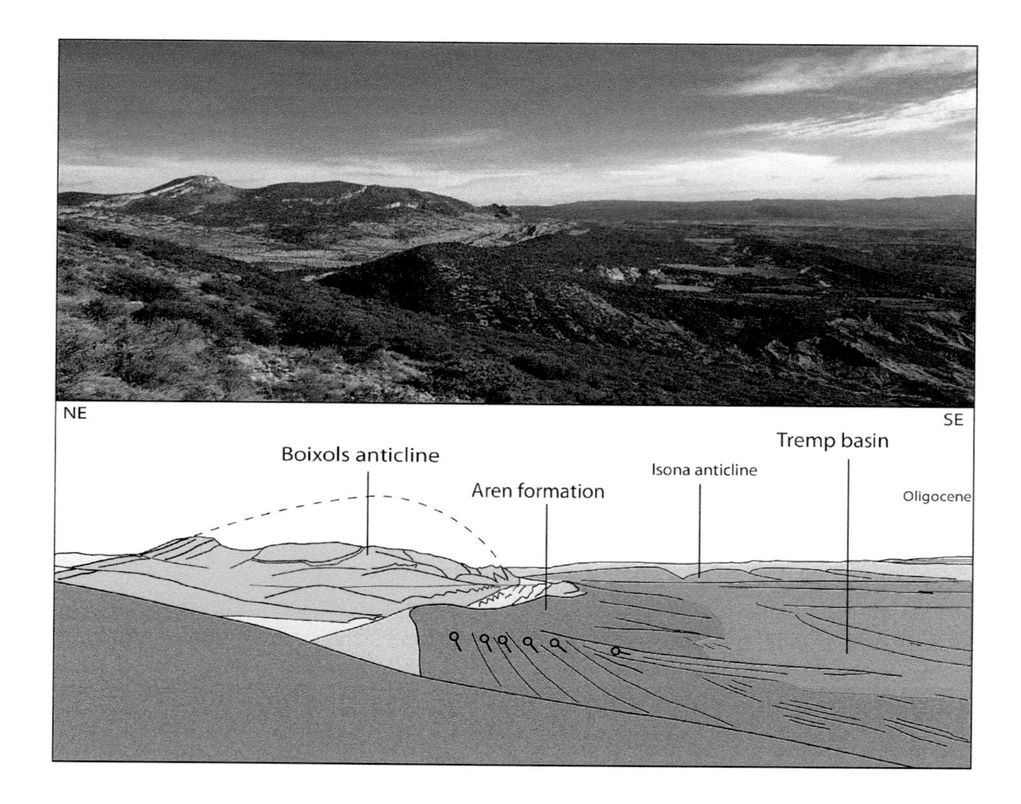

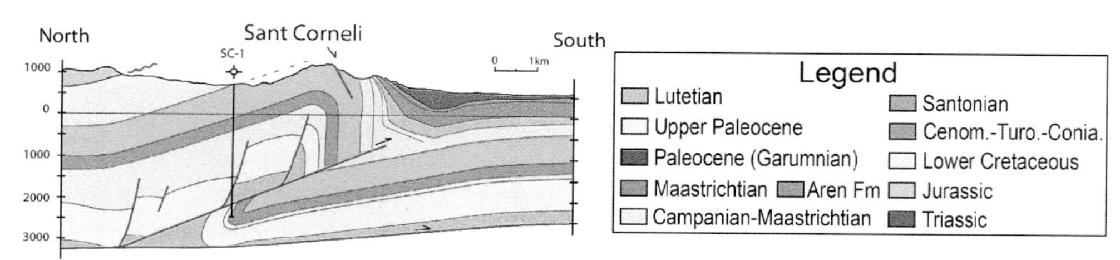

Figure 2.13 Photo, interpreted panorama and cross-section of the Sant Corneli anticline (Tremp, southern Pyrenees) showing progressive unconformities (brown layers) at the front of the fold, truncated by a second unconformity (red layers) sealing the folding phase. The Sant Corneli fold is interpreted as a ramp propagation fold (see Chapter 5).

Source: Photo and interpretation by Romain Robert (2018)

four pieces constituting the tilted layers underneath the unconformity are connected two by two by lateral slides. The part representing the horizontal deposits over the unconformity is linked to the other four by a slide. The colors have been chosen to roughly respect the order of the international stratigraphic chart.

A valley is formed by a notch intersecting the five pieces and provides a view, from above, of the layers below the unconformity. The presence of this valley also establishes an order, ensuring that the parts are not interchangeable.

Figure 2.14 The unconformity at Minerve (Hérault, France). The strata forming the cliff date from the Eocene; those outcropping into the valley are from the early Paleozoic (Cambrian). Note the absence of "basic conglomerate" (see Fig. 2.1). The brown dotted line underlines the attitude of the Paleozoic strata below the unconformity. Even if the sedimentary hiatus is considerable here, on the order of 450 Ma, it is indeed the Hercynian unconformity fossilizing the Variscan tectonics. The slight dip observed in the limestone can be attributed to the bending of the European lithosphere at the front of the Pyrenees.

Source: Photo © Dominique Frizon de Lamotte

All the desired views can be obtained from this model, particularly the classic views for geologists, such as the sectional view and the map view (Fig. 2.15).

The second, more generic, model shows a stratum lying unconformably on a syncline/anticline pair. In detail, the folds are conical (Figs. 2.16 and 2.17). This model can initially illustrate the geometry of Paleozoic terrains in Normandy in contact with the sediments of the Paris basin. In this context, the red would represent the Icartian basement (Brioverian), the purple to orange are the Paleozoic strata and the yellow are the Mesozoic terrains. With this analogy, it must be emphasized that the stratigraphic colors (see the attached stratigraphic chart) are not correct. But nothing is stopping you from changing them when you come to producing your own model.

To close this chapter, Figure 2.18 shows an unconformity in the Southern Urals (Russian Confederation) that is quite similar to the one in Normandy between the "Brioverian" and the Lower Paleozoic. Tilted Ordovician strata can be seen lying unconformably on almost

Figure 2.15 Unconformity 1, the constructed model: side view and map view. Note: the map view clearly illustrates the "V" rule in the valley explained in Chapter 1. The yellow stratum lies unconformably on the red, blue and green strata, which have been tilted and eroded.

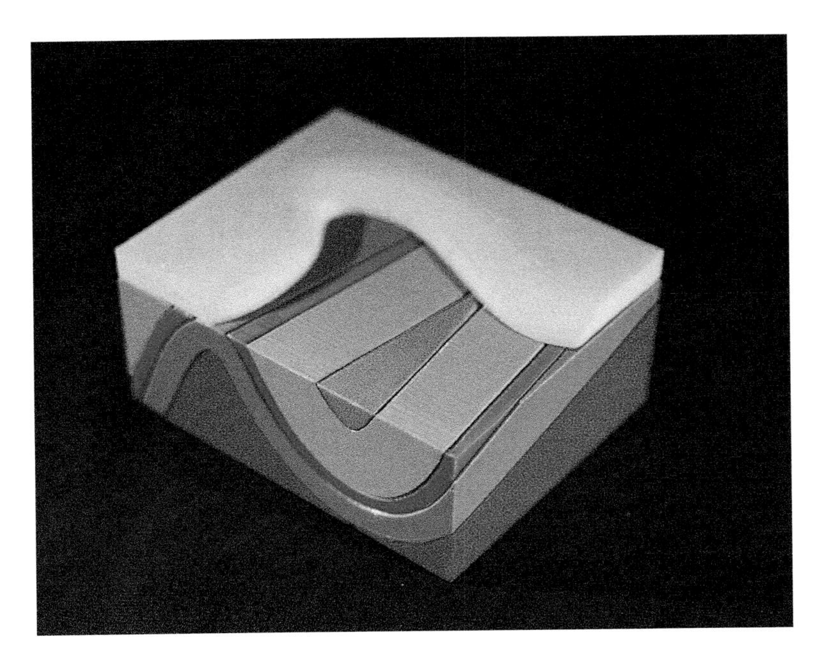

Figure 2.16 Unconformity 2: perspective view.

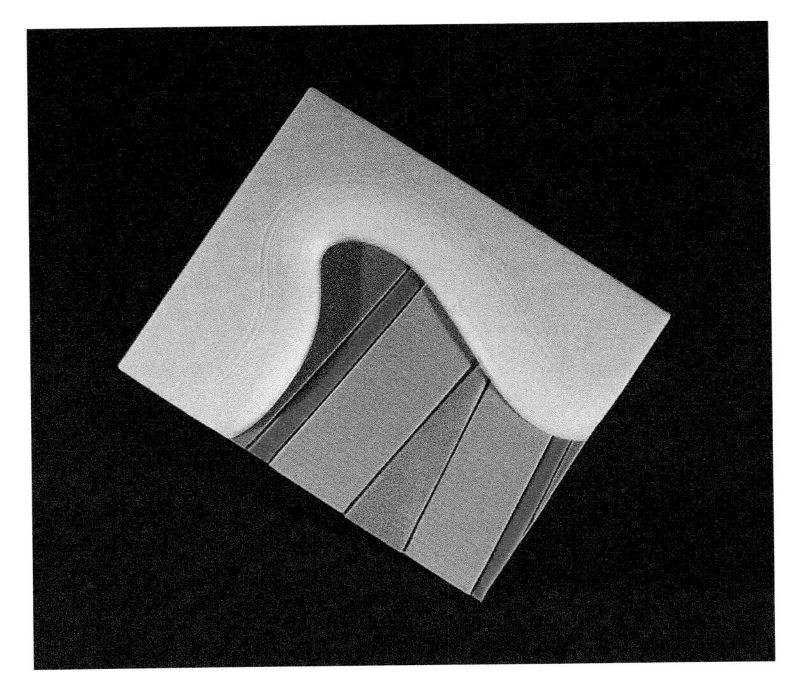

Figure 2.17 Unconformity 2: map view.

Figure 2.18 Ordovician conglomerates and sandstones (right) lying unconformably on alternating sandstone banks and upright Upper Proterozoic clay-rich banks (left). The Proterozoic basement and the Paleozoic cover are taken together in the late Paleozoic deformation, which gave rise to the formation of the Urals. The site is located in the southern Urals in the Republic of Bashkortostan (Russian Federation).

Source: Photo © Iskhak Farkhutdinov

vertical strata of the Upper Proterozoic (Riphean) time. This illustrates the relative nature of the basement concept. The deformed Proterozoic is the basement of the Paleozoic strata. The Urals range, which belongs to the family of Variscan chains, deforms both and thus forms the basement of new, younger basins outcropping on both sides of the chain.

Chapter 3

When strata fracture

Faults

3.1 Fracture theory (Mohr's circle and envelope)

Various forces act on the earth's rocks. These forces are generally classified into two categories: volume (i.e., gravitational) forces relating to the weight of the rocks and surface forces due to horizontal movements, and then ultimately plate tectonics. When forces are applied to surfaces they generate tectonic stresses. When the applied stress exceeds the strength of the rock, the rock is deformed and a change in volume and/or shape occurs. If this proceeds to failure, the deformation can result in faults. We will see that a simple tool, the Mohr circle, can be used to predict the orientation of these faults in relation to the forces acting on the rock. To do this, we first have to introduce the concept of stress, which is widely used by geologists. It should be noted that geologists work on the basis of constant volume, or rather they take it that volume changes generally take place before tectonic deformations occur (especially during sediment compaction). We will also assume the plane strain condition, that is the assumption that the deformation is in the plane that contains the axes of shortening and elongation without deformation in the third dimension.

We will consider a force F acting on a surface area a (Fig. 3.1). This force F can be decomposed into a normal force (F_n) and a tangential force (F_t) (Fig. 3.1B). The corresponding stress is: $\sigma = \frac{F}{a}$. When stress comprises a force acting over a surface area, it has the dimension of pressure [P_a]. This stress can be decomposed into two components: a normal component σ_n and a tangential component τ (Fig. 3.1C). They both act on surface area a' and their magnitudes are given by:

$$\sigma_n = \frac{F_n}{a'} = \frac{F \sin^2 \alpha}{a} = \sigma \sin^2 \alpha$$

$$\tau = \frac{F_t}{a'} = \sigma \sin \alpha \cos \alpha$$

In the convention used by geologists, normal stresses are considered positive when compressive and negative when tensional, while shear stresses are considered positive when dextral and negative when sinistral.

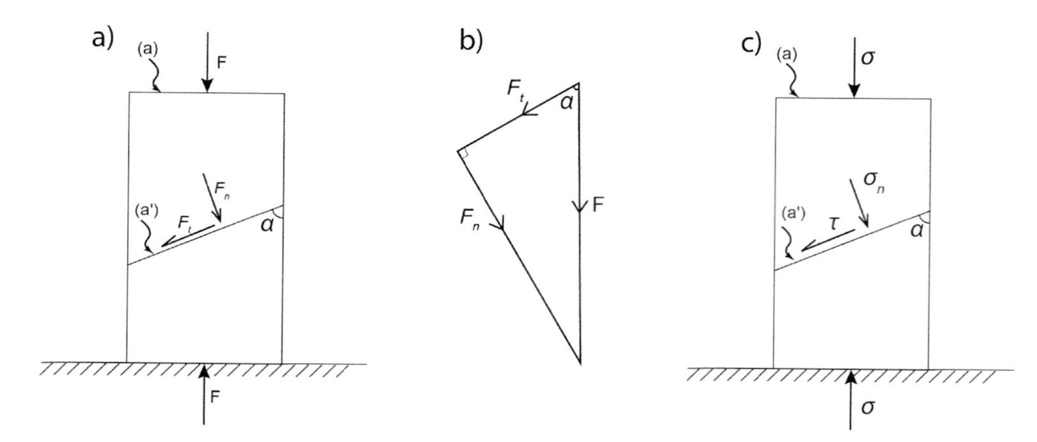

Figure 3.1 Left: an experimental arrangement showing a cylinder to which a vertical force F is applied. Centre: breakdown of the forces acting on the cylinder. Right: the breakdown of the stresses acting on the cylinder; a is the surface area to which force F is applied; a' is a surface area having an inclination of α with respect to F, on which F_n and F_t are separated.

3.1.1 *Mohr circle*

Geologists frequently encounter two problems:

1 Establishing the components of the stresses acting on a known orientation plane.
2 Finding the principal stresses when the stress components are only known in one plane.

Solving these problems in three dimensions is complicated. Nevertheless, in two dimensions a graphic solution can be found by constructing a "Mohr's circle." We will therefore consider the problem in two dimensions. Readers who would like to consider the problem in three dimensions are referred to the "Triaxial stress condition" insert.

We will consider a cylindrical block having a cross-sectional area 1 in any vertical plane passing through the axis of the cylinder. We define principal stress conditions σ_1 and σ_3 acting on each of the rectangular faces of these planes (Fig. 3.2). We consider an inclined surface at an angle α to the direction of σ_1 to be the maximum principal stress.

The normal and tangential stress components acting on this surface can then be calculated. The normal stress σ_n has two components due to σ_1 and σ_3, respectively:

$$\sigma_n = \sigma_n^{(1)} + \sigma_n^{(3)}$$
$$\sigma_n = \sigma_1 \cdot \sin^2 \alpha + \sigma_3 \cdot \cos^2 \alpha$$

The tangential stress also has two components due to τ_1 and τ_3, respectively:

$$\tau = \tau^{(1)} + \tau^{(3)}$$
$$\tau = \sigma_1 \cdot \sin\alpha \cdot \cos\alpha - \sigma_3 \cdot \sin\alpha \cdot \cos\alpha$$

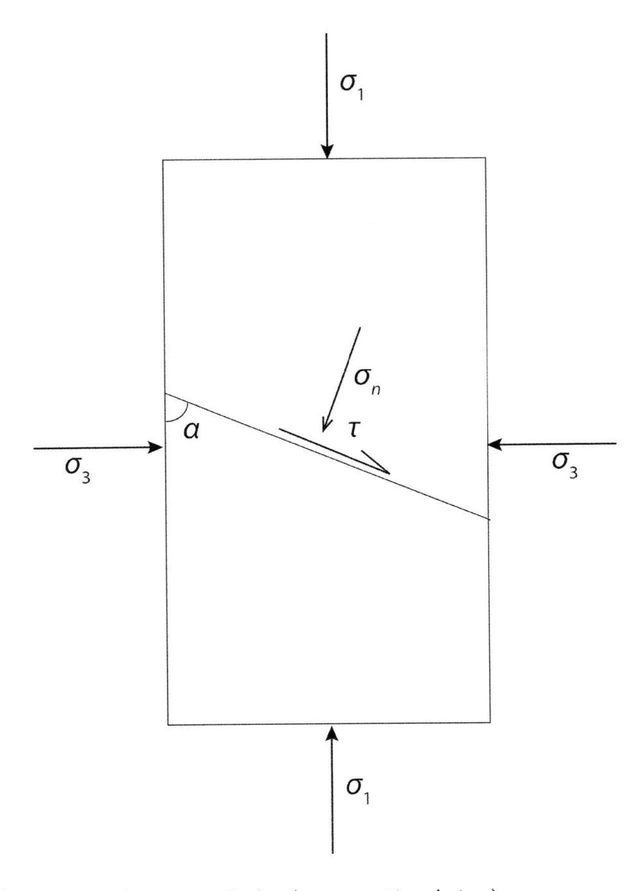

Figure 3.2 Biaxial stresses acting on a cylinder (cross-sectional view).

using the following trigonometrical relationships:

$$\sin\alpha \cdot \cos\alpha = \frac{1}{2}\sin 2\alpha$$

$$\sin^2\alpha = \frac{1}{2}(1 - \cos 2\alpha)$$

$$\cos^2\alpha = \frac{1}{2}(1 + \cos 2\alpha)$$

We can rewrite the normal and tangential stresses as:

$$\sigma_n = \frac{1}{2}(\sigma_1 + \sigma_3) - \frac{1}{2}(\sigma_1 - \sigma_3)\cos 2\alpha$$

$$\tau = \frac{1}{2}(\sigma_1 - \sigma_3)\sin 2\alpha$$

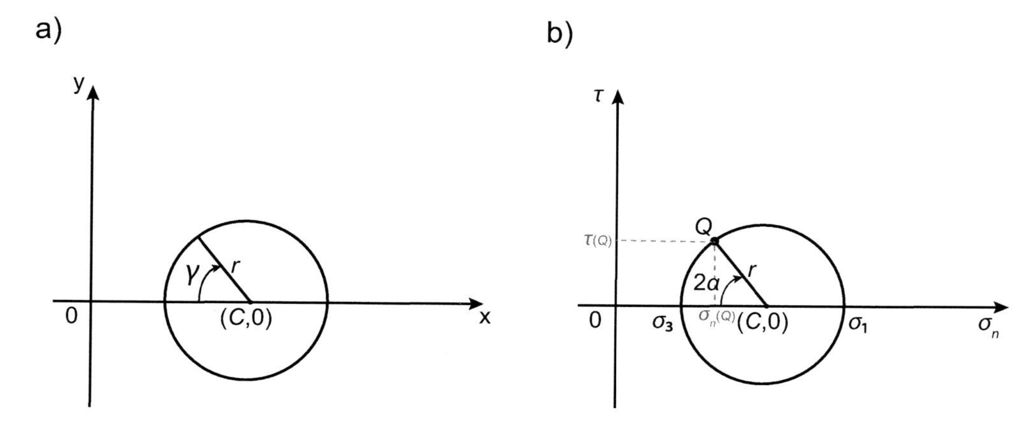

Figure 3.3 (a) Representation of the circle in the reference frame (0, *x*, *y*). (b) Representation of the biaxial stress condition.

These equations are of the same form as the parametric equations of a circle with a center having the coordinates (C; 0) and a radius r (Fig. 3.3a):

$$x = C - r\cos\gamma$$
$$y = r\sin\gamma$$

with the angle γ defined as: $2\alpha = \gamma$.

The principal biaxial stress condition can then be represented by a circle (Fig. 3.3b) in a system of axes (σ_n; τ). The values of the stresses acting in a plane having any orientation can therefore be read directly from the circle. If the plane is oriented at an angle α to the principal stress direction σ_1, then point Q is at an angle 2α from direction σ_1 on the circle.

Triaxial stress condition

For any facet corresponding to the element of a surface, we consider a vector **n** normal to this facet (outgoing by convention) and a vector **t** tangential to this facet. The orientation of the tangential vector is chosen so that the base {**n**, **t**} is direct (the angle between the two vectors is positive). In nature the general stress condition is represented in three dimensions (Fig. 3A_1). An index system is used to indicate in which direction and on which surface the stress acts. If we consider a cube in the Cartesian reference frame (*O*; **x**, **y**, **z**) the stress condition is defined by nine components. This is written as a 3 × 3 dimensional matrix called a stress tensor:

$$\sigma = \begin{pmatrix} \sigma_{xx} & \tau_{xy} & \tau_{xz} \\ \tau_{yx} & \sigma_{yy} & \tau_{yz} \\ \tau_{zx} & \tau_{zy} & \sigma_{zz} \end{pmatrix}$$

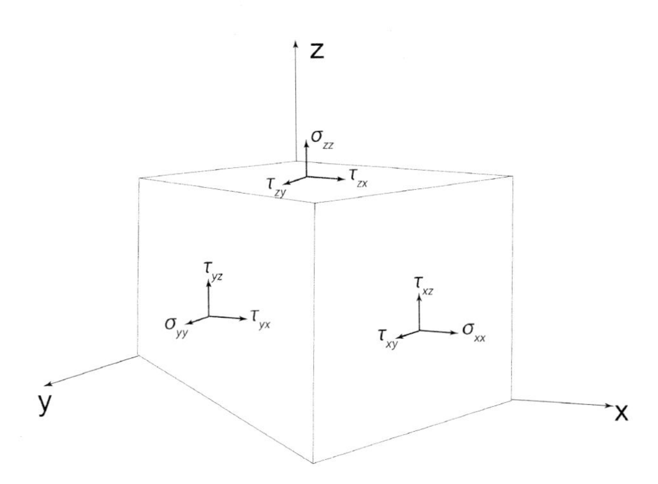

Figure 3A$_I$ Components of the stress condition acting on any volume.

with three normal stress components σ_{xx}, σ_{yy} and σ_{zz} and six tangential stress components τ_{ij}. For example, τ_{xy} is the tangential stress on normal surface x in shear direction y. The tensor is symmetrical if $\tau_{xy} = \tau_{yx}$, $\tau_{yz} = \tau_{zy}$ and $\tau_{xz} = \tau_{zx}$. The stress condition then includes six independent terms.

If the cube is rotated in any orientation the stress condition is preserved. However there are three specific directions in which the tangential or shear stresses are zero. The stress tensor is then diagonal and the components are called principal components. We note the principal stresses σ_1, σ_2 and σ_3 and the tensor is written:

$$\boldsymbol{\sigma} = \begin{pmatrix} \sigma_1 & 0 & 0 \\ 0 & \sigma_2 & 0 \\ 0 & 0 & \sigma_3 \end{pmatrix}$$

Whatever the principal stress condition may be, the mean stress is defined by:

$$\sigma_m = \frac{\sigma_1 + \sigma_2 + \sigma_3}{3}$$

For a fluid, the hydrostatic stress condition is defined as: $\sigma_1 = \sigma_2 = \sigma_3 = p$ and $\sigma_m = \frac{p}{3}$ where p represents the hydrostatic pressure. The mean stress tends to cause a change in the volume of the material without any change in shape.

On the other hand, we define the deviatoric stress tensor σ^d as:

$$\boldsymbol{\sigma} = \sigma_m \begin{pmatrix} 1 & 0 & 0 \\ 0 & 1 & 0 \\ 0 & 0 & 1 \end{pmatrix} + \boldsymbol{\sigma}^d$$

The deviatoric tensor expresses the change in shape in the material. In the principal frame of reference each principal stress is broken down into two terms: a mean stress and a deviatoric stress expressing the deviation from the mean. In direction 1 the total tensor is written as follows: $\sigma_1 = \sigma_m + \sigma^d_1$

For a triaxial test, this deviatoric stress tensor is written as follows:

$$\sigma^d = (\sigma_1 - \sigma_3)\begin{pmatrix} \frac{2}{3} & 0 & 0 \\ 0 & \frac{-1}{3} & 0 \\ 0 & 0 & \frac{-1}{3} \end{pmatrix}$$

3.1.2 *Mohr envelope*

The aim is now to define a strength criterion for a given rock. The experiment illustrated in Figure 3.5 provides an intuitive idea of this criterion. We will imagine a block that is rigid but of negligible mass in equilibrium on a rigid foundation. This is the interface between the two solids whose strength is under test. This is the shear test, well known in geotechnical engineering (Fig. 3.4a).

The tangential force T required to initiate slip is an increasing function of the normal force F_n applied to the top of the block. We therefore measure $\tau = \frac{T}{S}$ during displacement of the block of rock for constant $\sigma_n = \frac{F_n}{S}$, S being the basal surface area of the block (Fig. 3.4b). The test is then repeated by changing the value of the normal force. After a series of shear tests we obtain the maximum values for the tangential stresses required to break the rock for the different values of normal stress applied; these are the values denoted τ_{peak} in the diagram (Fig. 3.4b). These values lie on a straight line in the reference frame $\{\sigma_n, \tau\}$ (Fig. 3.4c). This linear relationship between the two forces is certainly the first solution that can be proposed to describe the onset of slip. This is the Mohr-Coulomb failure criterion, which is written as follows (Fig. 3.5):

$$\tau = \sigma_n \cdot \tan\varphi + c$$

It involves the two physical parameters specific to the rock in question: the friction angle φ [degrees] defined by the gradient of the straight line ($\mu = \tan\varphi$ is the friction coefficient widely used in geology) and the cohesion c [Pa] defined by the intersection between the Mohr envelope and the τ axis. In the general case this failure criterion is a straight line in the compression part but a curve in the tension part ($\sigma_3 < 0$).

We can use this strength criterion to determine the orientations of possible failure planes. In fact, when the circle is tangential to the Coulomb line, then the orientation of the failure

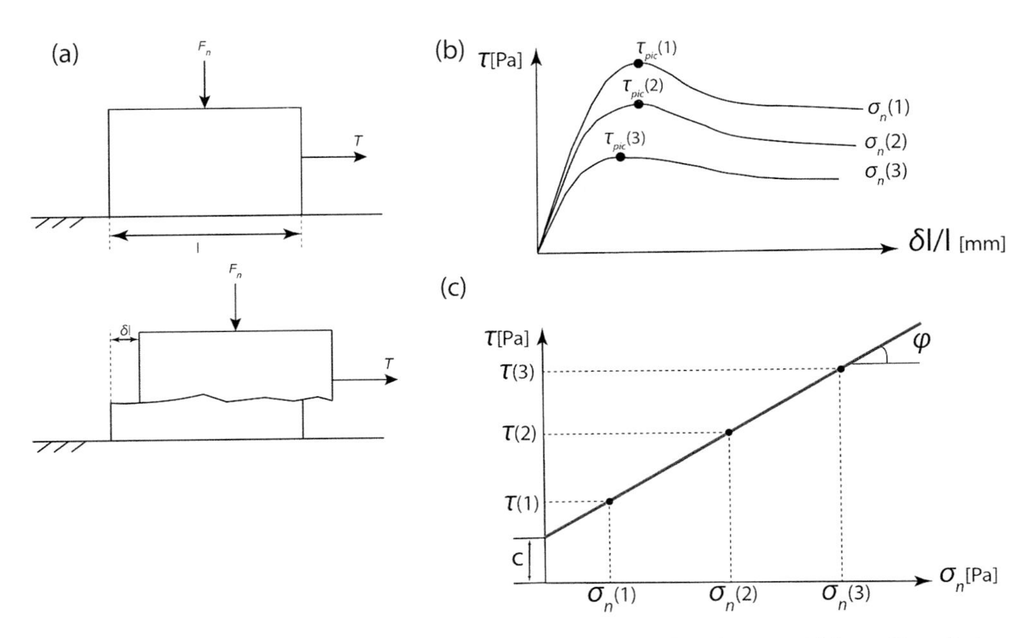

Figure 3.4 (a) Representation of a shear test with normal force F_n and tangential force T applied to the block of rock. (b) Variations in tangential stress in relation to the relative displacement of the block for different normal stress values. (c) Linear relationship between tangential stress τ and normal stress σ_n.

plane can be deduced as shown in Figure 3.5. This orientation is given by the following relationship:

$$\sin 2\theta = \frac{2\tau_c}{\sigma_1 - \sigma_3}$$

where τ_c is the tangential stress at failure and θ is the orientation of the potential fault plane.

It can also be deduced that: $\theta = \frac{\pi}{4} - \frac{\varphi}{2}$. This relationship links the orientation of the plane and the friction angle. Finally, we can see on this diagram that for a plane at 45° to the maximum principal stress, the maximum shear stress τ_{max} that the rock can withstand before failure is half the differential stress $(\sigma_1 - \sigma_3)$:

$$\tau_{max} = \frac{\sigma_1 - \sigma_3}{2}$$

For natural rocks, the Mohr envelope is only linear in a few special cases (sand). In general, the shape is that of a parabola whose axis of symmetry is the axis of the σ_n (Fig. 3.6). This shape explains the occurrence of special fractures such as tension cracks or compaction

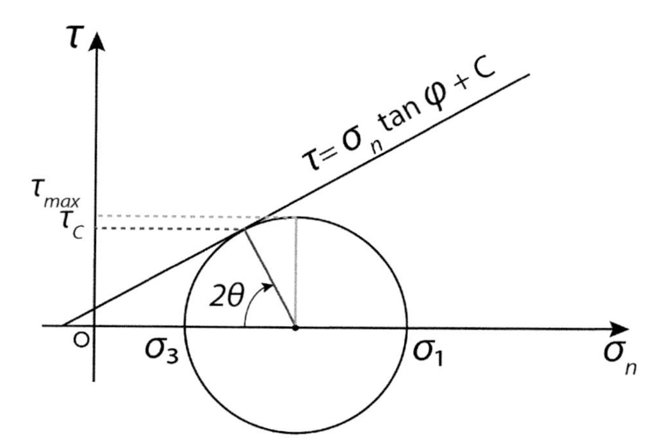

Figure 3.5 Determination of the orientation of a potential fault plane (in orange) and the maximum tangential stress on a rock for a plane at 45° to the direction of the principal stress σ_1.

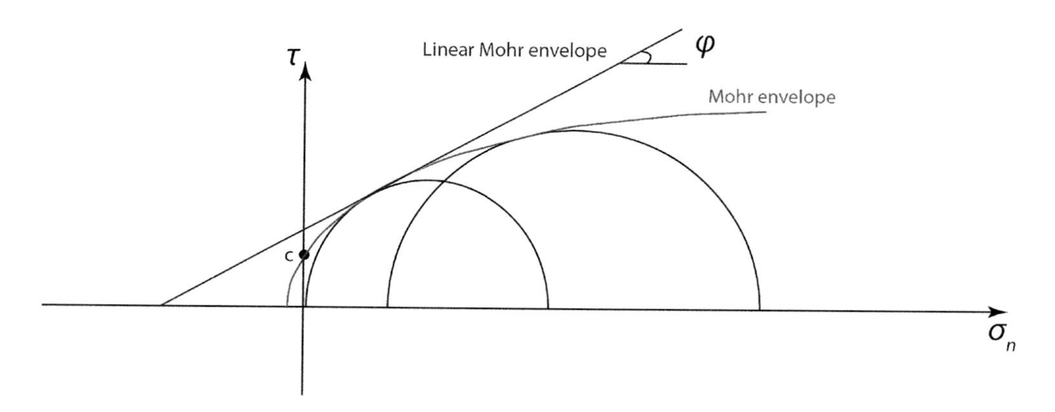

Figure 3.6 General appearance of the Mohr envelope as experimentally obtained for a rock. Note the significant deviation from the linear envelope.

bands (see the box "When rocks fracture in other ways"). It also explains why the dip of faults changes with depth, as we will see later.

When rocks fracture in other ways

Faults are the most spectacular and common features due to brittle deformation. However, rocks can break in another way without invoking shear planes. This applies to the "tension cracks" that are common in limestone rocks. In this case fracture of the rock occurs parallel to the principal stress σ_1. The fact that the tension gash is filled with a fluid that crystallized indicates that the opening is under tension ($\sigma_3 < 0$, in the geologists' sign convention). Thus in the Mohr space we are precisely at the point where the Mohr envelope intersects the axis of the σ_n (Fig. 3A$_2$).

Fractures can also develop parallel to the principal stress σ_3. The fracture then results from crushing of the grains constituting the rocks to form "compaction bands," thus reducing porosity. Other members of this family are "stylolites" and the early stages of the development of cleavage, sometimes called "fracture cleavage" to indicate that the rock breaks easily along this plane. On all these aspects, and others, we refer to the work by J. Mercier, P. Vergely and Y. Missenard (Dunod ed). The Mohr envelope can be closed on the right by an elliptical envelope representing the compaction failure limit, as shown in Fig. 3B$_2$.

Tension cracks or compaction bands can coexist with faults; this is even very common. To illustrate this type of combination, Figure 3C$_2$ shows the classic site of Terres Rouges in the Corbières, known to many geologists. This outcrop exposes silts and calcarenites at the base of the Cenozoic ("Vitrollian" facies, which is widespread in Provence and the eastern part of the Pyrenees).

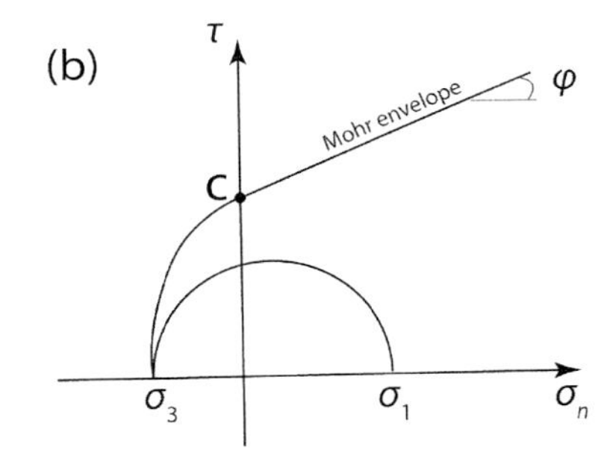

Figure 3A$_2$ (a) Tension crack and the directions of the associated principal stresses. (b) Mohr's circle corresponding to this tension crack. The photo is of Cretaceous limestone in the Pyrenees.

Source: Photo by Eduard Roca

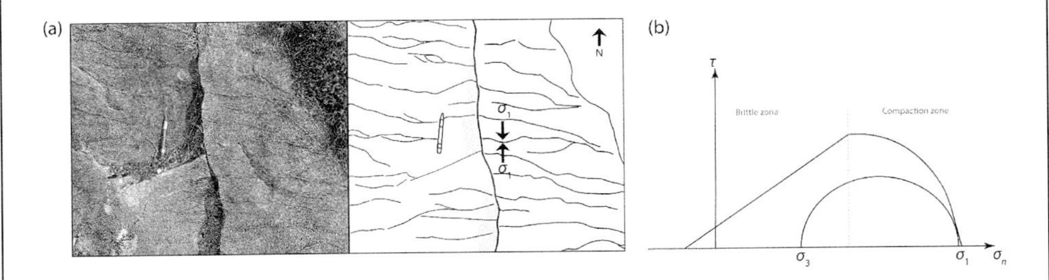

Figure 3B$_2$ (a) Compaction bands and directions of the associated principal stresses. (b) Mohr's circle corresponding to these compaction bands according to Risnes, 2001. The photo was taken in Upper Cretaceous sandstone near Tremp (Spain).

Source: Photo by Romain Robert

Figure 3C₂ (a) Cleavage associated with reverse faults and directions of the associated principal stresses. (b) Mohr's circle corresponding to the simultaneous development of these two structures, theoretical diagram without scale. The has been taken in the "Vitrollian" of the Terres Rouges site near the village of Tournissan.

Source: Photo by Dominique Frizon de Lamotte

Large horizontal X-shaped structures can be seen, indicating the presence of reverse faults. The maximum main compression direction σ_1 is parallel to the strata (slight dip toward the NW). These combined faults give us an angle 2θ, which is here about 40°. This low value shows that we are very far to the left on the σ_n axis abscissa in the Mohr space (Fig. $3C_2$). This is evidence of the role of fluids in causing this shift in the circle to the left. This role of fluids is indicated by the abundance of calcite on the fault planes.

There is also cleavage perpendicular to the strata. Like faults, this results from tectonic compaction parallel to the strata, known as "layer parallel shortening" (LPS). In the upper layer (Fig. $3C_2$), faulting and cleavage coexist. In the lower layer, only the cleavage is expressed and is very dense. This underlines the role of lithology (the nature of rocks) in the expression of deformation.

3.1.3 *The role of fluids*

Most rocks are porous to a greater or lesser extent and contain a fluid (water, oil, natural gas, etc.). The effect of the presence of this fluid is to reduce the effect of the principal compressive stresses on the rock. We have seen earlier that the normal and tangential stresses are expressed respectively, as follows:

$$\sigma_n = \frac{1}{2}(\sigma_1 + \sigma_3) - \frac{1}{2}(\sigma_1 - \sigma_3)\cos 2\alpha$$

$$\tau = \frac{1}{2}(\sigma_1 - \sigma_3)\sin 2\alpha$$

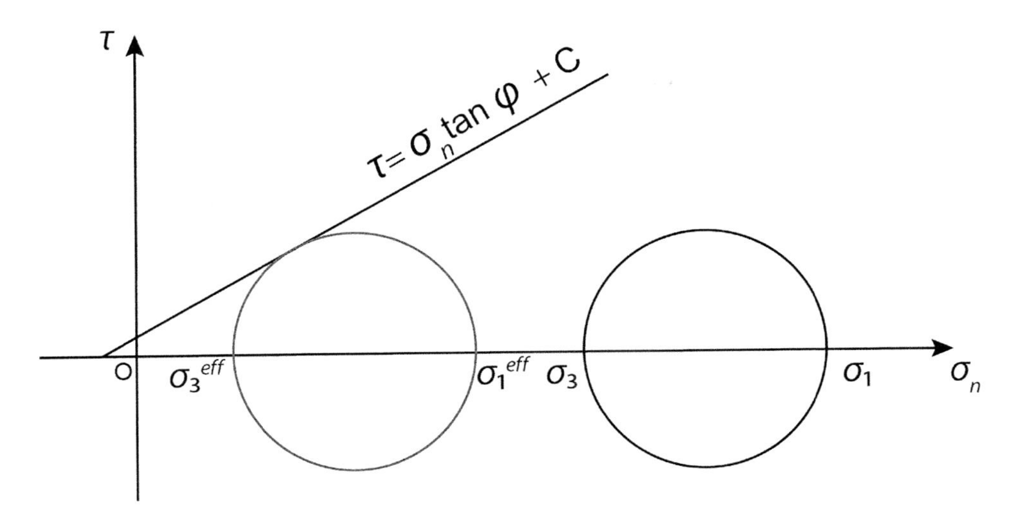

Figure 3.7 Displacement of the Mohr's circle in the presence of fluids. The effective stresses are denoted σ_i^{eff} (i = 1, 2 or 3).

If we substitute ($\sigma_1 - p$) for σ_1 and ($\sigma_3 - p$) for σ_3, where p represents the fluid pressure, then these relationships become:

$$\sigma_n = \frac{1}{2}(\sigma_1 + \sigma_3) - \frac{1}{2}(\sigma_1 - \sigma_3)\cos 2\alpha - p$$

$$\tau = \frac{1}{2}(\sigma_1 - \sigma_3)\sin 2\alpha$$

The effect of the presence of fluid is therefore to reduce the normal stresses of a p component and to leave the values of tangential stresses unchanged. In terms of the Mohr's circle, this effect shifts the circle to the left along the axis of normal stresses. The rock is then weakened since it reaches the Coulomb criterion at lower stresses (Fig. 3.7).

3.2 The main categories of faults

Theory (Mohr's circle, see section 3.1.1) tells us that a fault develops at an angle of Θ with respect to the principal compressive stress σ1. The geologist's main interest is to determine the tectonic regime in which the fault develops. This can be extensional (σ1 vertical), compressive (σ1 horizontal, σ3 vertical) or strike-slipping (σ1 horizontal, σ2 vertical). The expected faults are respectively called normal faults in extension mode, reverse faults in compression mode or strike-slip faults in strike-slip mode (Fig. 3.8). Considering the behavior of the rocks (their rheology), angle Θ is theoretically always $\leq 45°$. As a result, with reference to the horizontal, normal faults tend to be at a high angle (on average 60°) while reverse faults are at a low angle (on average 30°). This has only been verified in the upper parts of the crust and in the absence of previous deformation, that is, if the faults are newly formed.

Establishment of these relationships between fault geometry and tectonic regime is due to E. M. Anderson and sometimes referred to as the "Andersonian" system. It should be noted that, theoretically, two faults develop simultaneously on either side of the main compressive stress σ1. These are referred to as conjugate faults. Nevertheless, by releasing stresses, the development of a fault will very often inhibit the development of its conjugate.

In general, the Mohr-Coulomb envelope (Fig. 3.6) is asymptotic for high stresses. As a result, faults tend to flatten at depth and can even become horizontal if they intersect a level with a low coefficient of friction, which the fault can link to. The level in question

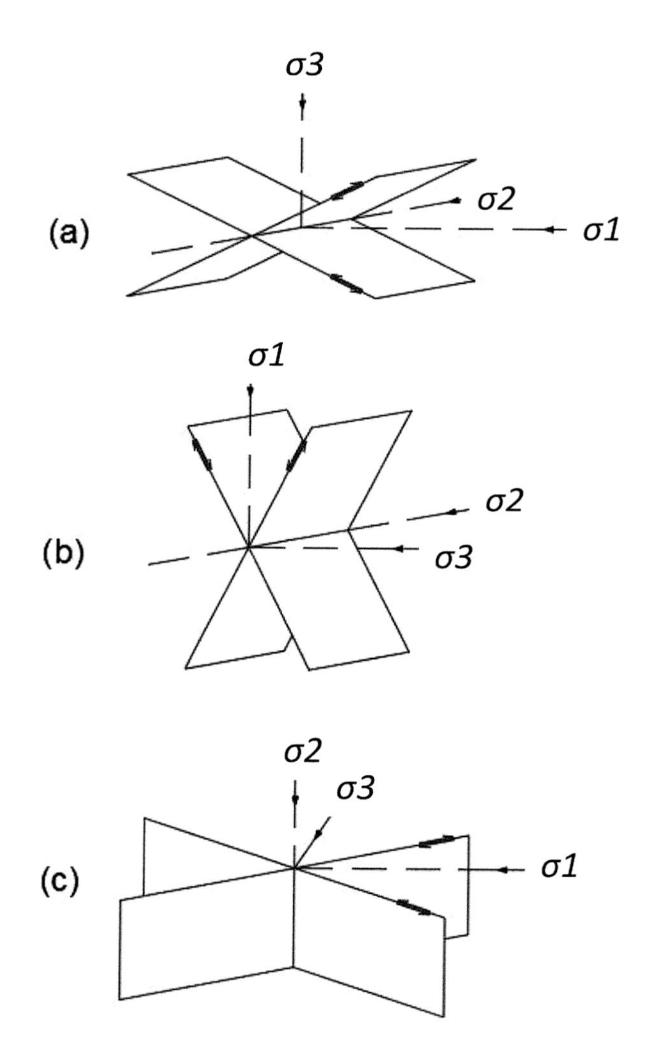

Figure 3.8 Types of faults and tectonic regime: (a) compressive regime, (b) extensive regime, (c) slip regime.

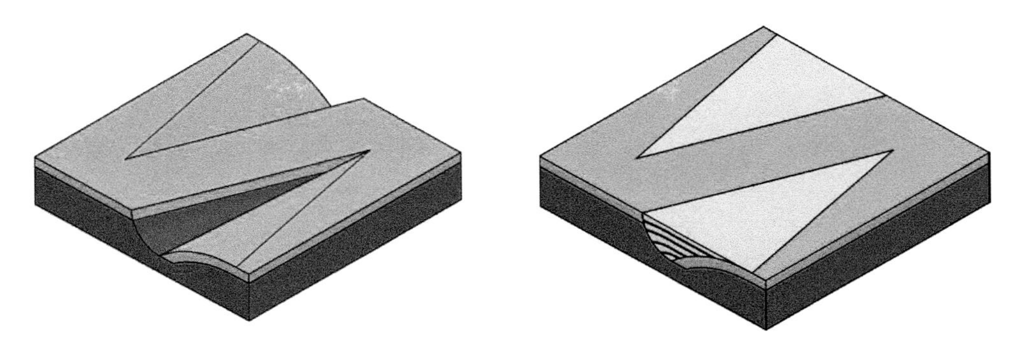

Figure 3.9 Listric and roll-over fault, on the left without sediment fill, on the right with fan-shaped sediment fill (progressive unconformities).

is then referred to as the detachment level (or simply detachment) and the fault becomes a listric fault with a typical curved shape (Fig. 3.9). This can be observed on various scales, from outcrop to the lithospheric scale. In a sedimentary basin the strata most likely to act as the detachment level are primarily evaporites, especially salt, but also clay or marl levels.

On the scale of the continental crust the same geometry can be seen when a fault reaches the ductile lower crust and continues there. The effect of the change in the dip of a fault causes the strata on the hanging side of the fault to fold down to form a so-called "roll-over" fold (Fig. 3.9) when in an extension regime. In compression mode, the rules governing the ways in which folds and faults interact are more complex, and these will be the subject of Chapters 4 and 5.

If there is a strong structural heritage, faults will preferentially use preexisting discontinuities. This can lead to geometries that are very different from those predicted by theory for a uniform environment. A spectacular case is that of detachment faults. These are very flat normal faults that sometimes reuse previous compression structures. In this case, they can develop near the surface. Their geometry may also result from rotation during evolution of the deformation (Fig. 3.10) or from a combination of both.

The question of structural heritage in a particular region is a major topic. Inheritance can in fact impose unusual geometries on faults. This is the problem of tectonic inversion. Thus, a normal fault can be reactivated as a reverse fault while keeping its steep dip. But in the same context a reverse fault can also be seen to ignore the previous normal fault and recut it.

Generally speaking, and regardless of its dip, a normal fault is "subtractive," that is it brings young strata into contact with older strata by "subtracting" part of the initial pile (Fig. 3.11). Conversely, a reverse fault is "additive" because it brings old strata into contact with younger ones by increasing the thickness of the initial pile (Fig. 3.11). It will therefore be understood that the cumulative effect of normal faults will lead to thinning of the crust while the effects of reverse faults will lead to thickening of the crust.

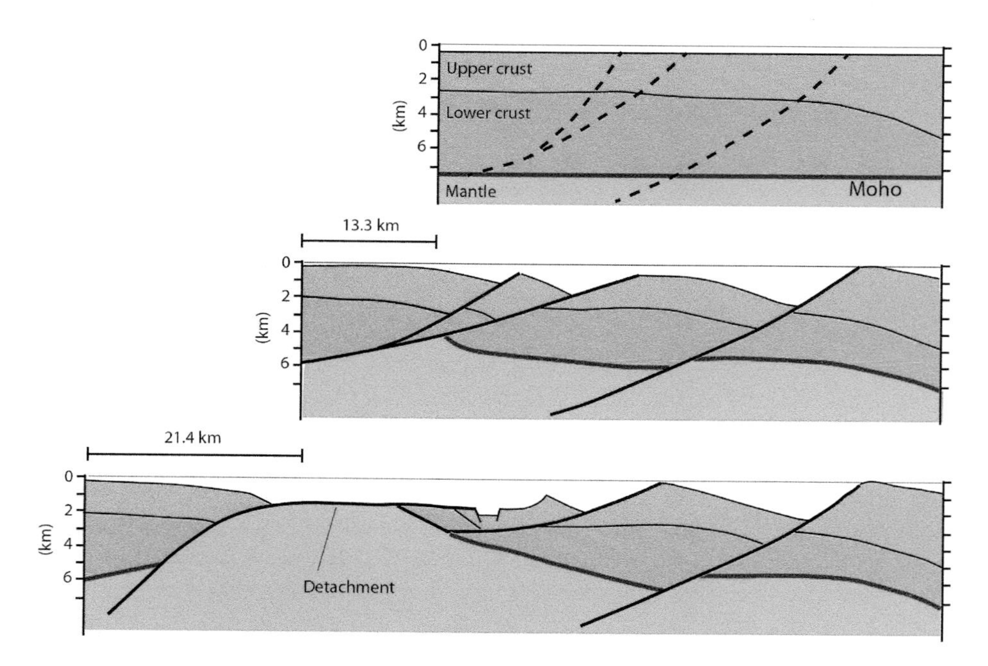

Figure 3.10 Conceptual diagram showing exposure of the mantle along a detachment fault. Note the rotation of faults over time. Note the upward convexity of the detachment fault as opposed to a listric fault, which is downwardly convex (Fig. 3.9).

Source: After Gianreto Manatchal

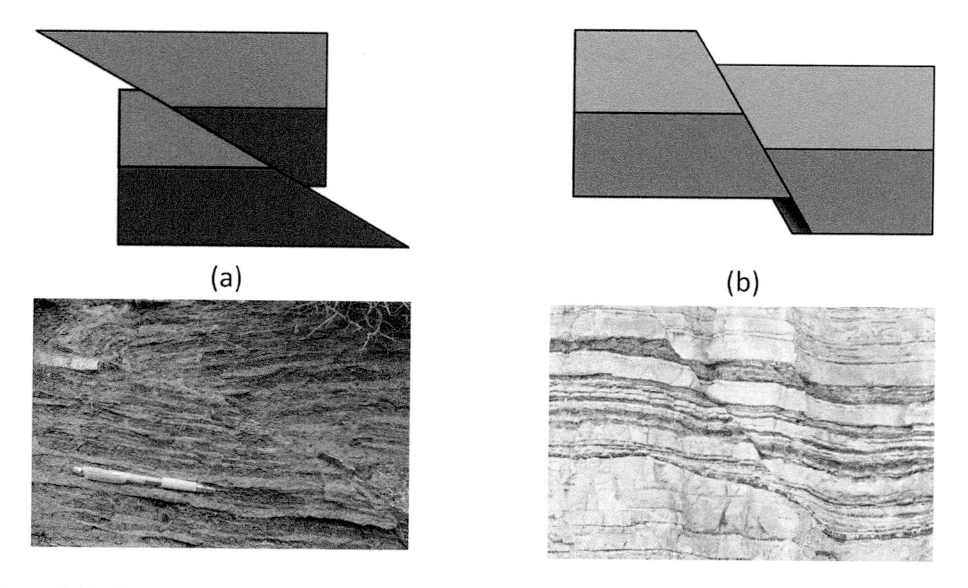

Figure 3.11 Diagrams and photos illustrating the additive nature of a reverse fault (a) and the subtractive nature of a normal fault (b). The photos were taken in Iran in the Kermanshah region.

Source: (a) © D. Frizon de Lamotte and (b) Paul Maguire-Fotolia.com

3.3 The consequences of movement on faults

3.3.1 *Fault propagation*

When talking about faults it is very important to make a distinction between the propagation of the fault, that is the mobilized surface, and the slip that occurs there. The propagation is necessarily larger than the slip. This can be discussed on an instantaneous basis. For example, after an earthquake, a 50 km break and a few meters of slip may be seen. A geologist can also measure it in the finished state. The slip then corresponds to the cumulative movement, the result of several earthquakes. Analysis of the relationships between the propagation (P) of a fault and the slip (S) that occurs there is very instructive.

3.3.2 *The concept of P/S ratio and the deformation related to the passage of a fault*

When a fault propagates, deformation is considered to occur at the tip of the moving fault. This deforming zone corresponds to a "ductile deformation bubble"; it is attached to the tip of the fault with which it moves (Fig. 3.12). At every moment during propagation of the fault, it is thus recut by the bubble, and the deformation is fixed on either side of the fault or on only one side in the case of detachment on a rigid substratum. The P/S ratio can be used to mathematically link three independent parameters: the slip on the fault plane (S), the propagation of the fault (P) and the elongation undergone by the strata during passage of the fault (ε), i. e. the deformation.

In the very simple case of a horizontal fault, such that the deformation is only transferred to the hanging wall of the fault (Fig. 3.13), we have:

$P = \mathrm{Li}$, where Li represents the initial length

and: $\varepsilon = \dfrac{\mathrm{Li} - \mathrm{Lf}}{\mathrm{Lf}} = \dfrac{\mathrm{Lf}}{\mathrm{Li}} - 1 = -\dfrac{\mathrm{S}}{\mathrm{P}}$

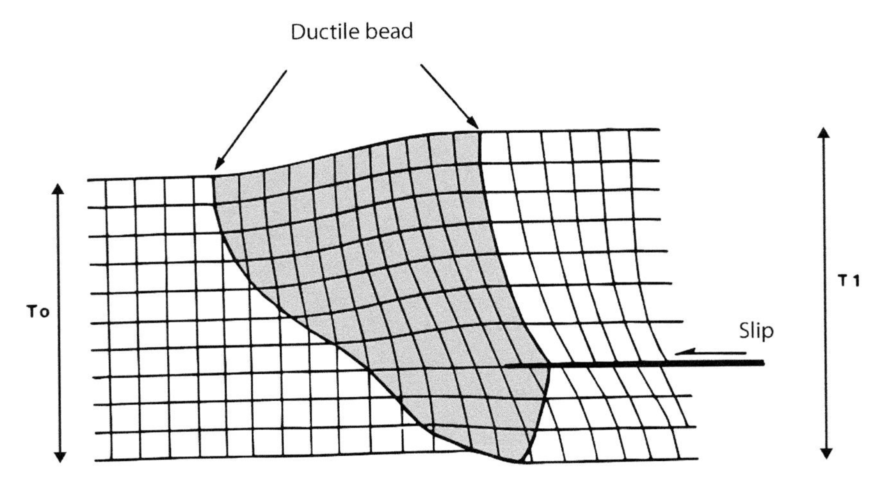

Figure 3.12 Illustration of the concept of a "ductile bubble" associated with the propagation of a fault.

Source: Drawing from Buil, 2002

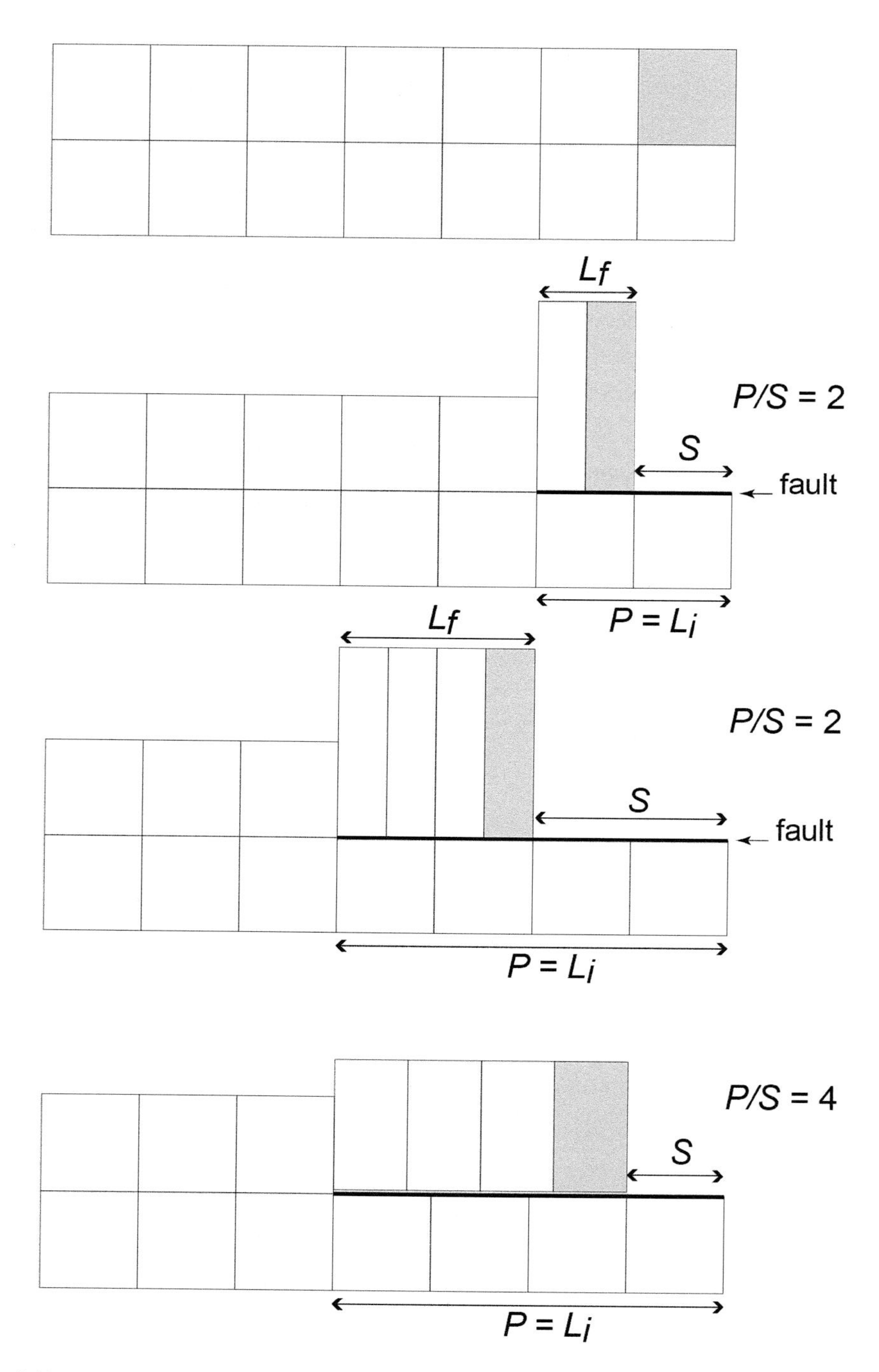

Figure 3.13 Relationship between the propagation of a fault, slip and deformation recorded in the hanging wall of the fault in the case of horizontal separation.

If we consider only the relative deformation ε_r measured parallel to the fault we have:

$$\varepsilon r = \frac{Lf}{Li} = 1 - \frac{S}{P}$$

$$\text{then: } \frac{P}{S} = \frac{1}{1 - \varepsilon r}$$

We can therefore see that the deformation ε_r is inversely proportional to the value of the P/S ratio. This means that the greater the propagation of a fault in relation to the slip, the less the associated deformation is strong.

3.3.3 *About cleavage*

This may seem quite theoretical, but through these very simple considerations it is possible to understand observations that are on the face of it intriguing. Thus, on the northern front of the Pyrenees, in the famous Corbières region, cleavage is seen to develop perpendicular to the strata (Fig. 3.14). This cleavage is indicative of horizontal tectonic compaction leading to the development of vertical movement. These are in fact dissolution planes marked by insoluble oxides and give rise to brittleness in the rocks along this anisotropic surface. This is unusual because, in general, *cleavage* does not develop under such superficial conditions (certainly less than 1000 m of burial in this case). We interpret the development of this cleavage as the result of several circumstances. The affected material had a very low initial anisotropy (it consists of highly bioturbated silts), so the deformation can be recorded immediately without the need to obliterate a previous fabric. In addition, the absence of an

Figure 3.14 Photo illustrating the development of cleavage perpendicular to the strata in the "Vitrollian" (Montian) silts of the Corbières. This structure illustrates shortening parallel to the strata.

Source: Photo © Pascale Leturmy

effective detachment level inhibits the propagation of a fault at the base of the sedimentary cover. Conditions are therefore favorable for strong deformation related to this difficulty in propagation of the detachment. Estimates suggest that in the case of Corbières the shortening parallel to the strata is of the order of 20%, which is equivalent to a P/S ratio of 5 (Fig. 3.14) (see also the insert: "When rocks break in other ways").

3.4 Elemental structures resulting from combinations of faults

Faults are rarely isolated structures but develop in a fault zone composed of different spatially connected segments. Each segment is separated from the next by an unfaulted connection, a kind of bridge that allows the fault zone to be crossed without intersecting a single one (Fig. 3.15). During progressive deformation leading to the creation of a fault zone, the general trend is for segments to be connected during propagation of the fault zone. In general, faults combine to give rise to structures on variable scales.

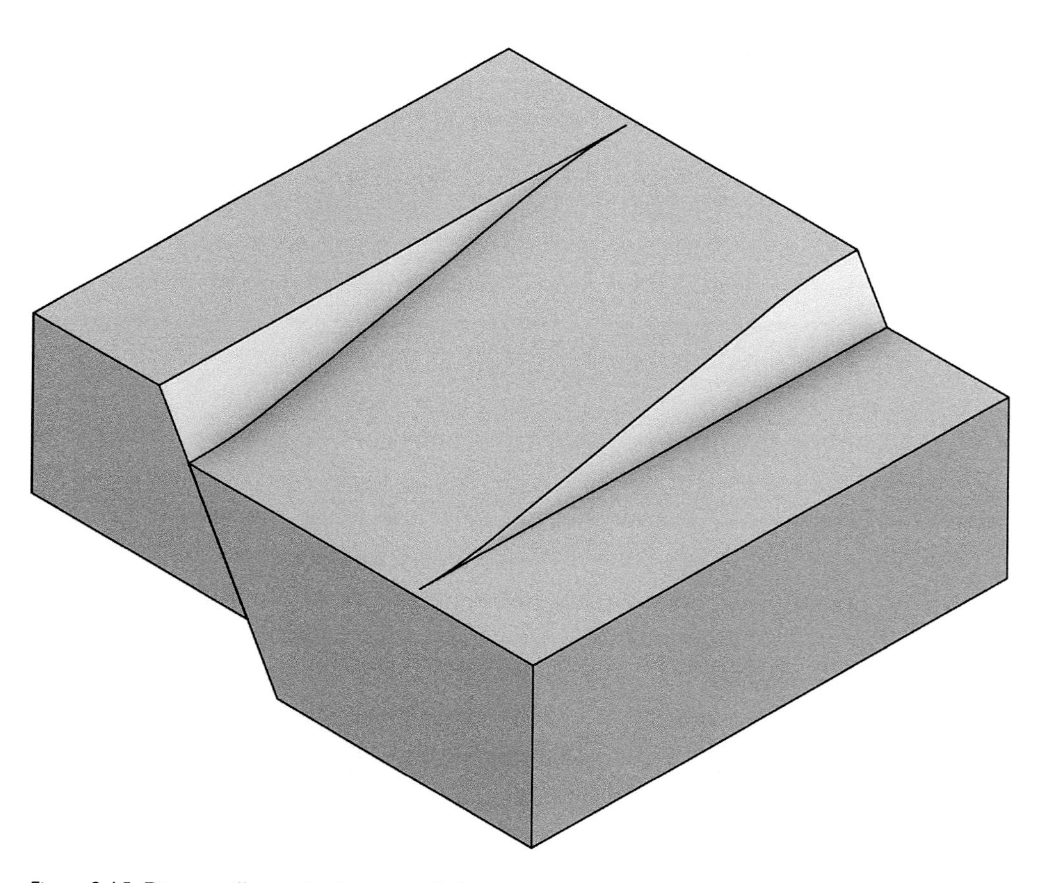

Figure 3.15 Diagram illustrating how two faults can alternate in space. The diagram is constructed so that the amount of extension is the same in any section perpendicular to the two faults.

3.4.1 Combination of faults in an extensional regime

In an extensional regime, faults combine to form rift systems, a consequence of the process called rifting, which consists of thinning the lithosphere until eventually it ruptures. Rift systems are composed of downthrown blocks (grabens) separated by raised blocks (horsts). They can have different geometries (called rifting modes) reflecting different geodynamic environments. Thus, a rift system can be wide (composed of several juxtaposed narrow grabens) or narrow (thinning is concentrated on a single structure) (Fig. 3.16). The rifting mode depends on several factors such as the thickness of the lithosphere, its thermal state and the brittleness due to structural heritage.

Another important aspect is the symmetry of the system. Rifts do not always resemble the simple picture shown in Figure 3.14, with a plane of symmetry along the system axis and steeply dipping faults. Asymmetry, which we have already encountered with regard to listric faults, is generally expressed by the development of a detachment fault on one side of the system (Fig. 3.10). This results in a lower plate below the detachment of an upper plate above the detachment. This inherited geometry will lead to asymmetry in the fill and will finally dictate the nature of tectonic inversion when the rift closes. It should be noted that the development of a detachment fault that cuts across the entire crust and allows the lithospheric mantle to be exposed requires a minimum amount of coupling between the different strata constituting the rift system (sediments, upper crust, lower crust, lithospheric mantle).

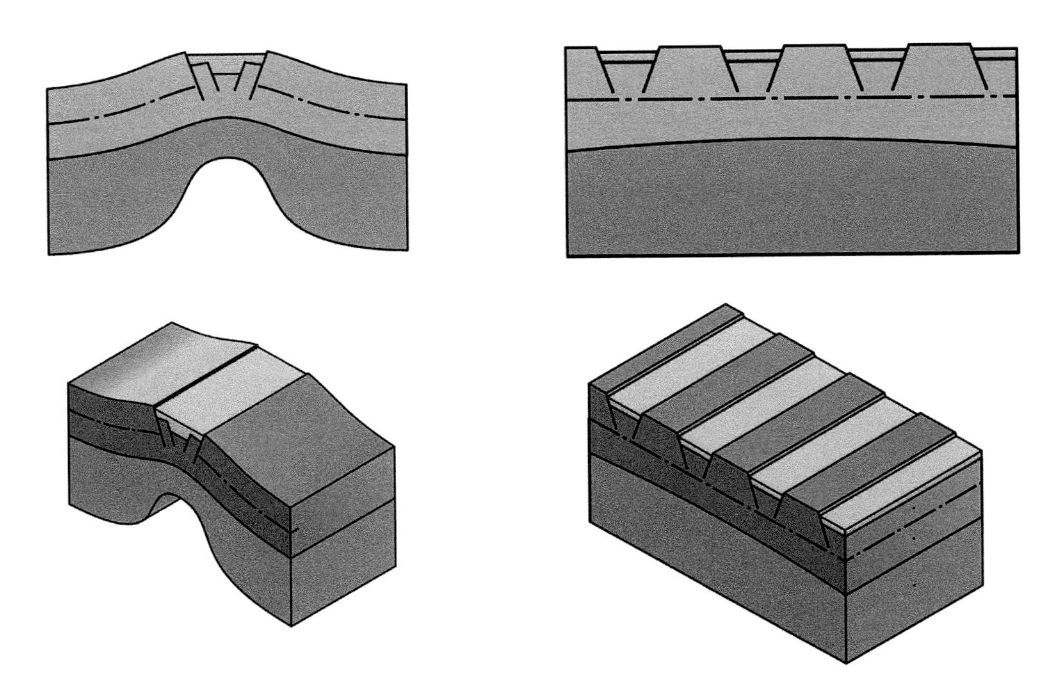

Figure 3.16 Concept of "narrow rift" (left) and "wide rift" (right). The yellow color represents sedimentary fill, the gray color represents continental crust and the green color the subcontinental lithospheric mantle.

This coupling is not present in the initial stages of rifting because the ductile lower crust isolates the upper crust from the mantle. Amplification of the deformation will lead to thinning of the lower crust to the point of allowing a fault to pass through the whole and couple the crust and mantle.

3.4.2 *Combination of faults in a compressional regime*

In compression mode, the faults also combine to form fold-and-thrust belts. As their name suggests, these structures intimately combine faults and folds. This topic will be discussed in detail in the following Chapters 4 and 5. However, we can mention here the concept of a "pop-up," which is a structure comparable to the "horst" in extensional tectonics. It is a plateau bounded by reverse faults on both sides (Fig. 3.17). This type of structure develops especially when the basement is involved in deformation.

3.5 Printable 3D models of faults and associations of faults

To illustrate the concept of a fault using 3D models, we intend to make use of two advantages of these models. The first is to enable us to understand how they evolve over time by accessing the third dimension. In fact, in geology, observing how a structure ends up also tells us how it has evolved over time. We will therefore put forward a model consisting of two listric faults arranged head to toe. The development of two roll-over folds in opposite directions from two distinct nucleation points can be seen (Fig. 3.18). The model is constructed in such a way that the extension is the same in all planes perpendicular to the two faults. Movement on the faults opens up a space for sedimentation that can be filled and then sealed by "post-rift" sediments.

The second advantage of our 3D models is that they can be used to play around with faults. This is enabled by a "dovetail" device, which is common in carpentry, allowing parts to be assembled. The only difference is that in this case we want the parts to be able to slip. We have chosen an arrangement in which the fault is inclined at 45° with respect to the stratification so that both normal and reverse faults can be triggered (Fig. 3.19).

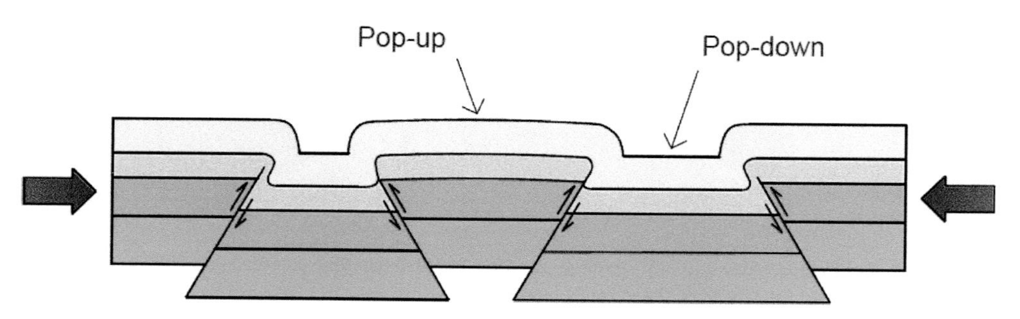

Figure 3.17 Diagram illustrating the concepts of "pop up" and "pop-down."

The arrangement in Figure 3.19 can be completed by a basement (representing the continental crust) into which the graben can be nested (Fig. 3.20). The graben, on the other hand, can be filled by a part representing the "syn-rift" sediments and ultimately covered by a final part representing post-rift sediments.

Figure 3.18 A 3D printed model combining two head-to-toe listric faults (left). The available space created can be filled by syntectonic sediments (yellow) and then covered by post-rift sediments (white). The part on the right is upside down. When turned over it fits onto the left-hand part.

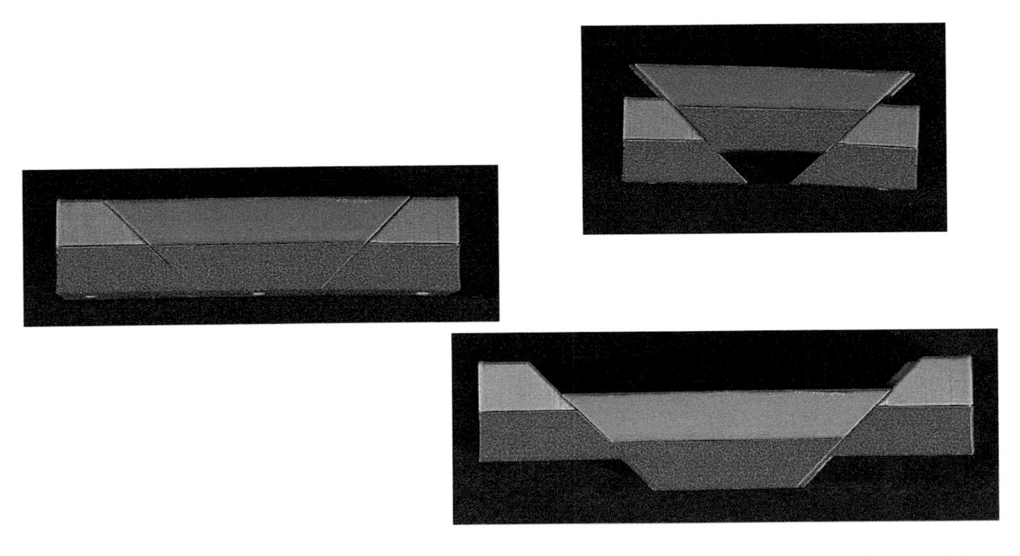

Figure 3.19 3D printed model enabling two conjugate faults to move to obtain either a "pop-up" (top left) or a "graben" (bottom right).

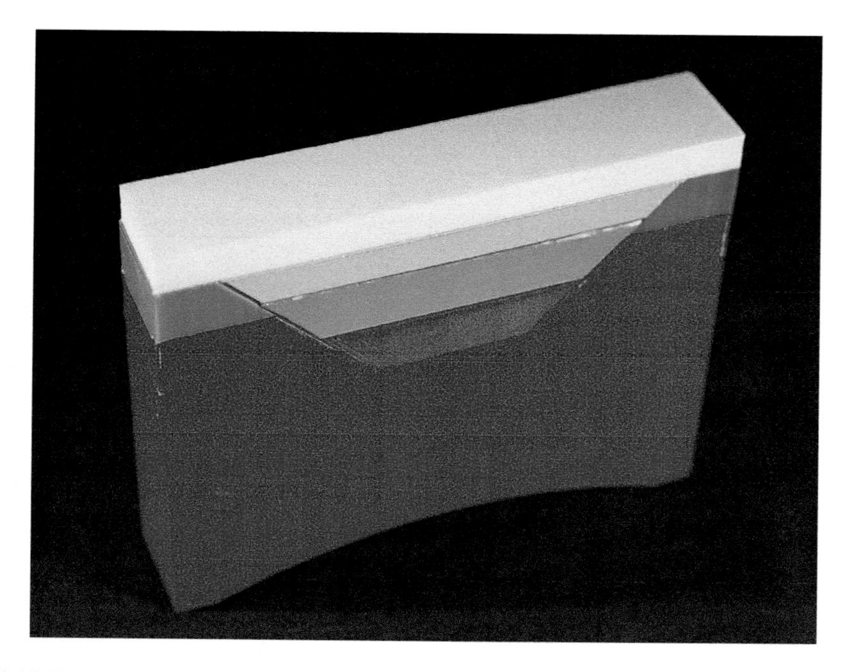

Figure 3.20 The model in Figure 3.18 supplemented by a basement, syn-rift fill (in green) and post-rift cover (in yellow). The model is constructed so that the "missing" volume in the crust base is exactly equal to that filled by the green piece representing the syn-rift sediments.

Chapter 4

When strata fold without breaking too much

Detachment folds

> As if the infinite were to have two levels: the folds of matter and the folds of the soul.
> (G. Deleuze, 1988)

4.1 Theory of folding

In mechanics, folding results from a process called buckling. It corresponds to the instability of an elastic structure (for example a beam). When this structure is subjected to compression, it will tend to bend and deform in a direction perpendicular to the compression axis (transition from a "compression state" to a "flexion state"). We have seen the compression state fossilized in the development of cleavage perpendicular to the strata in Figure 3.14. The flexion state is nothing more than a folded structure. Transposing this to a geological fold is not immediate for one simple reason. Fold development is not instantaneous; it extends over time, so that on geological time scales (horizontal shortening of the order of mm/year) the behavior of rocks is not elastic but viscous. Buckling is also possible in viscous media; it is described by Biot's law (1961):

$$\lambda = 2\,\pi\,h\,(\mu_1/\mu_2)^{\,1/3}$$

where λ is the wavelength of the fold, h is the thickness of the competent layer, η_1 and η_2 are the viscosities of the competent and less competent strata respectively (Fig. 4.1).

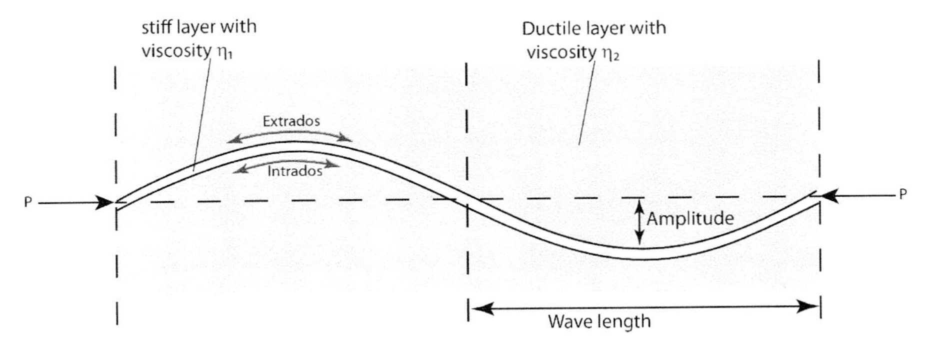

Figure 4.1 Viscous buckling according to Biot (1961), the diagram shows a 2-foot-thick layer of viscosity $\eta_1 = 10^{21}$ poise in a medium of viscosity $\eta_2 = 10^{18}$ poise. The lateral pressure exerted is P = 1450 psi. The length and thickness of the folded competent layer is preserved.

This equation is interesting because it shows that when folding is initiated, at least the wavelength (λ) depends mainly on the thickness (h) of the layer being folded and secondarily on the difference in viscosity ($\mu_1 \mu_2$) between the competent layer (the one that folds) and the surrounding environment. Conversely, it does not depend on the magnitude of the compressive forces applied. Once initiated the fold will be able to develop, that is grow in amplitude. The simplest idea is that amplification is acquired by reducing wavelength without changing the length of the limbs of the folds or, in other words, without changing the position adopted by the fold hinges since the initial incremental deformation (Fig. 4.1). But this is not without its difficulties. We will come back to this.

4.2 Some basic concepts relating to folds

In mechanics, bend tests are carried out on either vertical beams subjected to a force which is also vertical or horizontal beams supported at both ends and subjected to the effect of gravity. In the first case, deviation from the vertical may occur to the right or the left; it does not matter. In the second case, the beam will always bend with an upward concavity. In geology, strata are initially horizontal, and folds result from a shortening that is also horizontal. It is therefore useful to make a distinction between folds that are upwardly convex, referred to as anticlines, and those whose which are downwardly convex, referred to as synclines (Fig. 4.2). In principle, there is also a need to check that the youngest strata are at the center of the structure in synclines while the oldest are in the core of the anticlines.

Folding is always accompanied by deformation of the rocks making up the folded strata. In Figure 4.2, for example, choosing to keep the same radius of curvature inside and outside the folds requires thickening of the hinge with respect to the limbs of the fold. If we consider the classic case of strain plane condition, that is, the situation where everything happens in

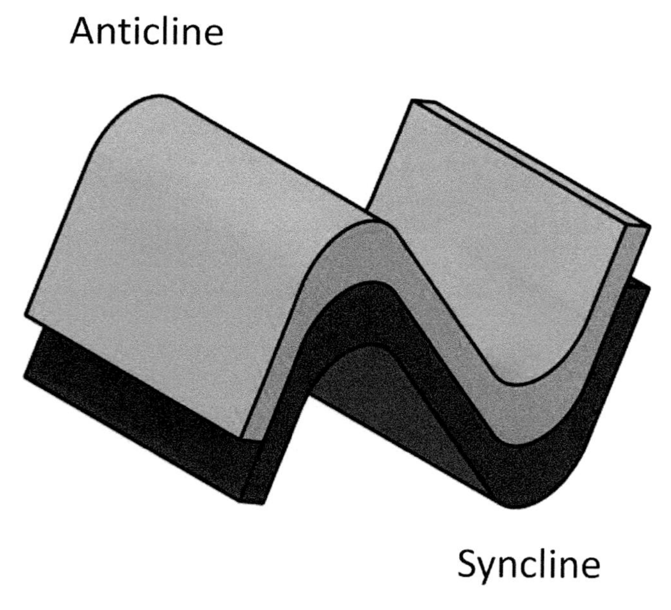

Figure 4.2 The concept of syncline and anticline.

the plane containing the axes of shortening and elongation, there is a transfer of material from the limbs to the hinges of the folds. If each layer is of constant thickness at the outset, then there will necessarily be thinning of the limbs of the fold. Geologists refer to these as similar folds.

However, during folding a significant part of the deformation is accommodated by slip between the strata. The thinner the strata, the more slip between the strata becomes predominant. In the case of a very thinly layered medium, this leads to the development of "kink" folds with sharp hinges and straight limbs (Fig. 4.3).

True *kink* folds, with sharp hinges, rarely develop in sedimentary rocks; they are more often found in rocks with very regular fine bedding such as shales or, more generally, rocks that have undergone metamorphism (Fig. 4.4c). In sedimentary rocks, folds are more commonly referred to as chevron, knee or even "chair" folds (Figs. 4.4a, b and d). The term chevron is purely geometric; knee suggests a bending mechanism with a fixed hinge. As for the term chair folding, to our knowledge it has only been used in the Franco-Belgian Ardenne massif. It evokes a particular type of chair that can be stacked on itself in large numbers. This is a particularly notable feature of kink folds (Fig. 4.3). Each layer is separated from the adjacent layer by a detachment level, or rather deformation is at the expense of a tight alternation of competent and less competent strata so that, in theory at least, their infinite stacking does not give rise to any geometrical problem (Fig. 4.3).

The case of *kink* folds provides an opportunity to transition to the concept of concentric folds. In this type of fold it is no longer the radius of curvature that is preserved but the center of curvature (Fig. 4.5). The consequence is that these folds are parallel folds. The

Figure 4.3 A 3D diagram illustrating the concept of "kink" folds; note the absence of a geometric limit to stacking in this type of structure.

Figure 4.4 Different sharply hinged folds: (a) narrowly hinged anticline fold in limestone-marl alternations in the eastern Zagros (Ilam region, Iran); note the slightly rounded nature of the hinge; (b) anticline fold in limestone-marl alternations on the Basque coast (Saint-Jean-de-Luz, France – the hinge is very clear and not thickened, it is a chevron or knee fold); (c) "true" kink folds affecting shales in the Kétama region (Central Rif, Morocco); (d) folds in alternating sandstones and clays in the Ardenne massif (Vireux region, France), referred to locally as "chair" folds.

Source: Photos © D. Frizon de Lamotte

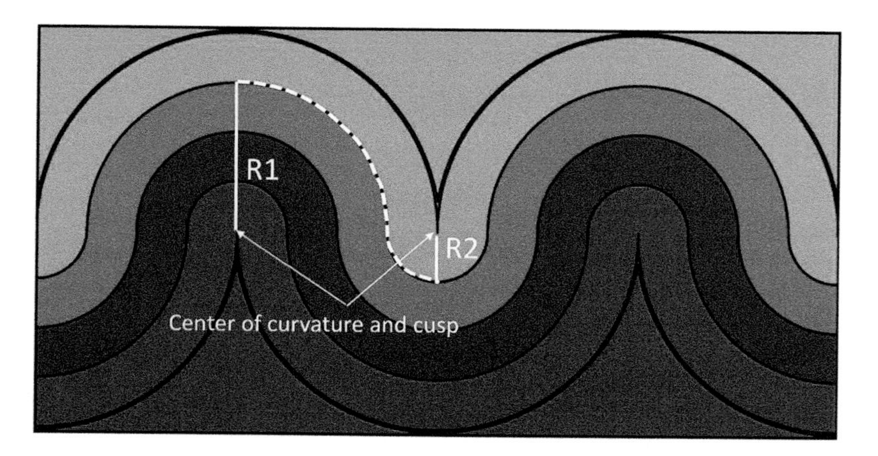

Figure 4.5 Top, concentric fold pattern. We can check that for any layer $L = (R1 \times \pi/2) + (R2 \times \pi/2) = \text{const.}$ This type of folding preserves the lengths of the initial strata until the center of curvature is reached. Layer thickness is also preserved at every position in the fold.

thickness of the strata is retained regardless of their position in the fold (flank or hinge). If again we consider a context of strain plane condition, there is necessarily a transfer of material from the inside (the intrados) to the outside (the extrados) of the hinge (Fig. 4.1). In section the alternation of anticlines and synclines has the result that the length of a layer is preserved between a syncline hinge and an anticline hinge. Layer length and thickness are therefore stored in a sedimentary sequence subjected to concentric folding (Fig. 4.5). In some cases, cleavage is seen to develop in association with the folds. Study of the relationships between strata and cleavage planes provides important information, as detailed in the insert relating to this.

Relationships between folds and cleavage

As we have seen in Chapter 3, cleavage corresponds to main deformation planes developing perpendicularly to $\sigma 1$. It can appear during the initial stages of deformation, even before folding begins to be expressed, or at the end of deformation when the folds have been completed. In the former case, the only one we will consider here, cleavage is initially perpendicular to the sedimentary strata (Fig. Aa).

During folding the cleavage is tilted with the strata and then takes on a fan shape, that is, the cleavage planes converge toward the axial surface of the fold (Fig. Ab). In this configuration the line of intersection between stratification and cleavage is parallel to the axis of the fold.

We also saw earlier in Chapter 4 that in some cases, particularly when there are fine alternations between competent and less competent strata, folding is accompanied by slip between the strata that concentrates in the least competent strata. This slip between the strata causes the cleavage planes in the least competent strata to shear while the cleavage planes in the competent strata remain more or less perpendicular to the strata. The cleavage in the least competent strata tends to parallel the axial surface of the fold. This configuration of cleavage planes is called cleavage refraction. The relationships between the cleavage planes and the stratification are then dependent on position in the fold (Fig. Ac).

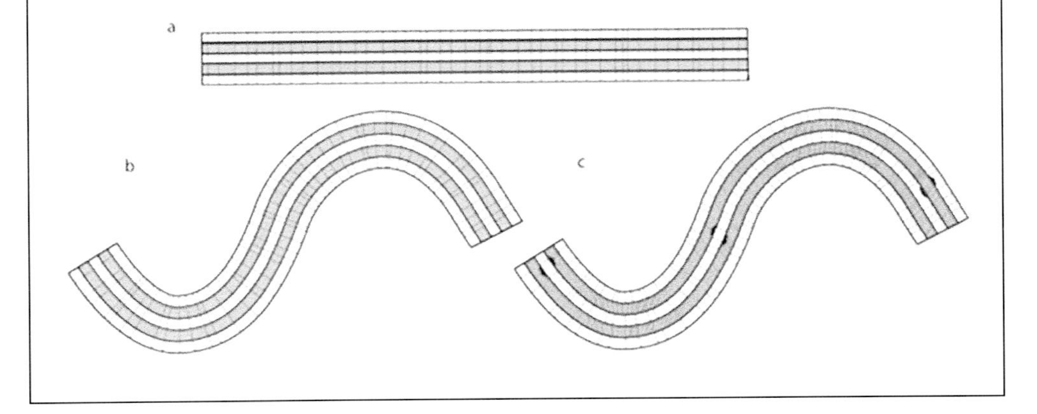

However, there is an obvious limitation to this type of folding, illustrated in Figure 4.5. Unlike *kink* folds it is impossible to stack them over a large thickness. Indeed, the fact that all the folds have the same center of curvature means that this center must also be a cusp beyond which the style of deformation changes drastically. We can therefore understand the need for an effective detachment level if this type of fold is to develop. Better still, geometry even makes an upper and a lower detachment level necessary, that is at the top and bottom of the set of folds in question (Fig. 4.6).

The bottom detachment level is absolutely essential. Most frequently it lies at the interface between the basement and the cover. This is the case, for example, in the Jura, the historical home of the detachment fold concept. In this case, it is located within the Triassic evaporites (Fig. 4.7). In the Jura, the sedimentary pile is thin (around 1500 m) so there is no upper detachment level, or rather, it was formed by the atmosphere at the time of folding. On the

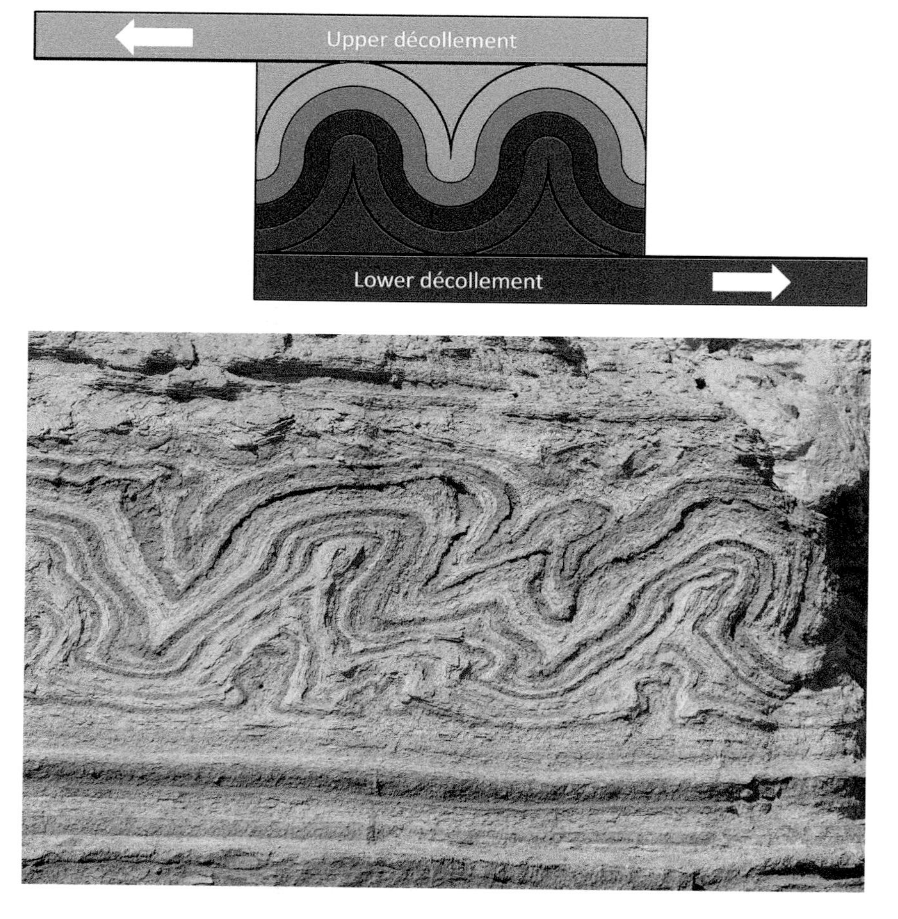

Figure 4.6 Top: diagram illustrating the concept of lower and upper décollement separating a set of isopach folds (drawing inspired by Dalhstrom, 1970, modified after Molinaro, 2004).

Bottom: photo of a set of folds bounded by detachment at the top and bottom.
Source: Dead Sea region, photo © J.C. Ringenbach

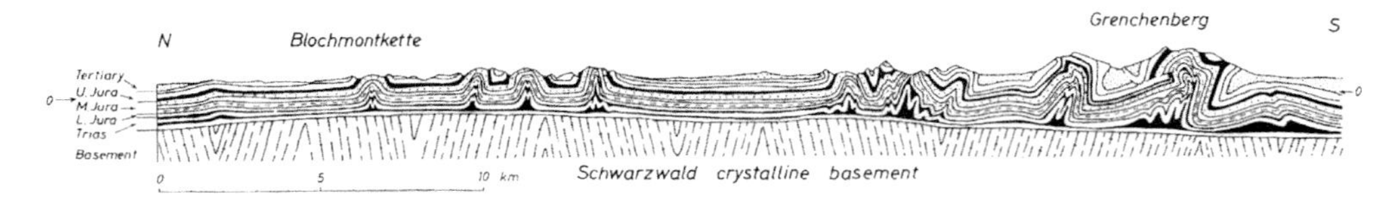

Figure 4.7 Historic section through the Jura (Buxtorf [1916] in Goguel [1952]) illustrating the concepts of basement, cover and detachment fold. The detachment here is ensured by the evaporites of the Middle or Upper Triassic (Keuper). In detail, the fold at the southern end of the section is more complicated than a simple detachment fold. The folds shown in the basement do not correspond to real structures; they indicate diagrammatically the existence of a previous orogenic cycle (the Variscan orogenesis). Using seismic technology modern studies have made it possible to see enormous Permian troughs under the Mesozoic.

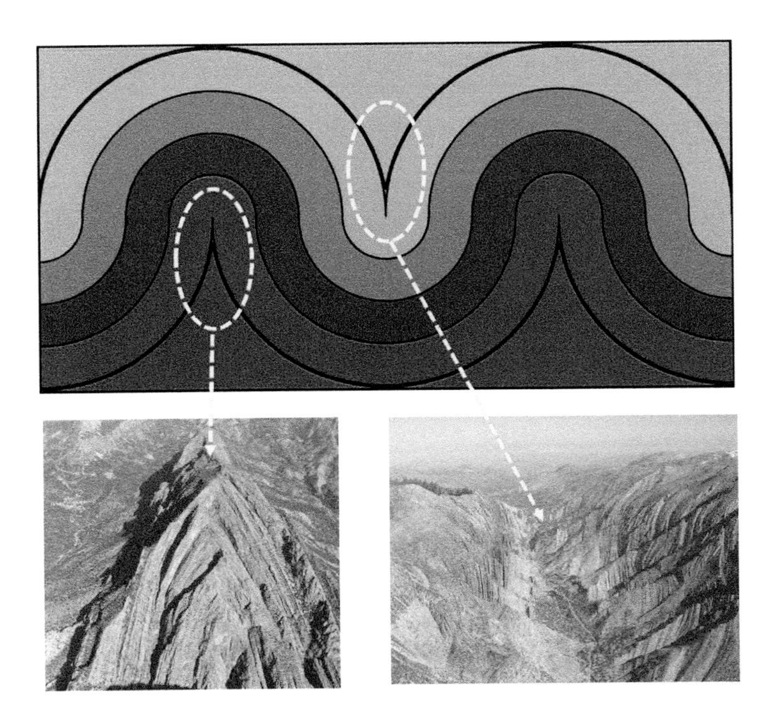

Figure 4.8 Illustration of the cusps in the core of an anticline and syncline in a concentric fold system in the Central Zagros (Izeh region, Iran). The upper and lower detachments are located in Miocene and Cretaceous age groups, respectively. Both are intermediate detachments in the stratigraphic pile, which is several thousand meters thick.

Source: Photos © D. Frizon de Lamotte

other hand, in a chain such as the Zagros (Iran), where the sedimentary pile is much thicker (up to 10,000 m and even more), sets of detached folds can be seen to be bounded by a detachment level at top and bottom. In practice, the Zagros has many levels of detachment in addition to the basal detachment located in the Hormuz saline formation at the boundary between the Precambrian and Cambrian. Figure 4.8 illustrates the basal cusps in the clays of the Kajdumi formation (Lower Cretaceous) and those at the top in the evaporites of the Gasharan formation (Miocene).

4.3 Horizontal shortening, excess surface area and detachment depth: toward the kinematics of detachment folds

We will now try to examine the consequences of the geometry of the detachment folds and understand how they can shed light on their kinematics, that is, their successive geometries over time. This is a matter of suggesting stages for the changes in shape to obtain the final structure from initially horizontal strata, which is what geologists assume. This approach will lead us to offer 3D models that incorporate the time dimension and take into account the concept of the balancing of geological sections (see the box on this topic).

What is a balanced cross-section?

The term *balanced cross-section* is widely used in the geological literature but is not necessarily completely clear to everyone. What is it about? A section is said to be balanced if the following conditions are met:

1 Competent strata (limestones, sandstones, etc.) retain their lengths and thicknesses between the initial and final states.
2 Less competent strata (salt, evaporites, clays, etc.) retain their surface areas between the initial and final states.
3 A geologically credible kinematic path links the initial state and the final state.

The first two conditions are based on the classical geologists' assumption of constant volume deformation. This assumption is reasonable. Indeed, it has been shown that in the main sedimentary rocks major changes in volume precede the development of folds. It is a compaction process, whether related to rock burial (sediment compaction) or horizontal shortening (tectonic compaction). In the case of tectonic compaction prior to folding, the layer-parallel shortening responsible for it can amount to 20%. It is therefore not in any way trivial and can considerably alter the physical properties of rocks (porosity and permeability in particular). It is during this early stage that small structures such as those mentioned in Chapter 3 (cleavage, compaction bands; see the box on this topic) develop.

The last condition is the essence of our book. In fact, to define a kinematic path between an initial state and a final state is a matter of defining intermediate sections and therefore being able to produce a 3D model. Here again the third dimension is the representation of change over time. All our models are therefore balanced since we have complied with the conservation criteria.

To begin with we will consider 2D, that is a section with a constant surface area (Fig. 4.9). We will call Li the initial length of the strata involved in a fold, Lf the final length of the folded pile, Xi the initial thickness of the detachment level and Xf the final thickness of this

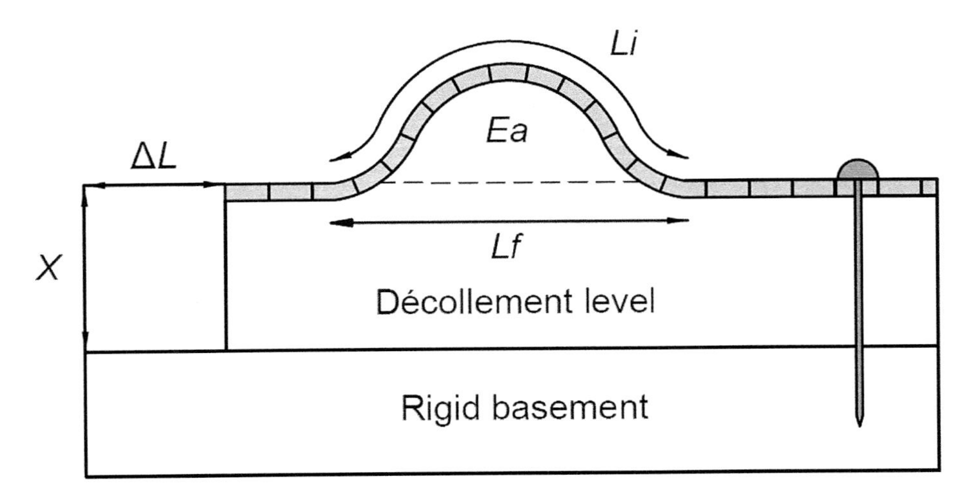

Figure 4.9 The theoretical principle of the excess surface area method initially proposed by Goguel (1952). This method would appear to make it possible to determine the depth of the basement (not involved in deformation) from the simple measurements of surface objects. However, it does not apply in common cases because in general, and contrary to what is drawn, the value of X (depth of the basement) is not fixed (see text). *Li*: initial length; *Lf*: final length; ΔL: shortening; *Ea*: excess surface area.

level. In an actual structure the width of the structure considered (*Lf*) is always measurable. The initial length of the strata (*Li*) is also easily obtained, sometimes at the cost of a few assumptions about how eroded strata in the core of the fold can be reconstructed. This length measurement is used to calculate the shortening ($\Delta L = Li - Lf$). Here we will also assume that the deformation is planar: there is no elongation or shortening in the direction perpendicular to the observed section. Any horizontal shortening will then be expressed by the creation of relief. Following Jean Goguel (1952, *Traité de Tectonique* [Treatise on Tectonics]), we give the name of "excess surface area" (*Ea*) to the surface area bounded at the top by the outline of the base of a given layer and at the bottom by the horizontal line marking the position of this same layer before folding. The excess surface area (*Se*) is therefore the apparent value of the structural relief created with reference to the base of a given layer (Fig. 4.9).

There is obviously a link between horizontal shortening ΔL and the structural relief measured by the excess surface area (*Ea*). Jean Goguel proposed a very simple relationship by postulating that the thickness of the detachment level under the synclines was constant: $X = \Delta L/Ea$. In fact this method was designed to find the value of X, which was considered to be fixed ($X = Xf = Xi$), the other quantities being measurable (*Li*, *Lf*) or easily calculable (*Ea*). However, geological experience shows us that in general this assumption is not true. When the detachment level is a salt level, it can even disappear completely from under the synclines by expulsion to adjacent anticlines. That is why we said earlier that the structural relief was only an apparent relief. Let us look at this question in more detail by exploring various configurations.

In mountain chains such as the Jura, where there is an effective detachment level, box folds are frequently present. These are anticlinal or synclinal folds with a horizontal top and base and vertical lateral limbs. Like cylindrical folds, they are parallel folds (Fig. 4.10).

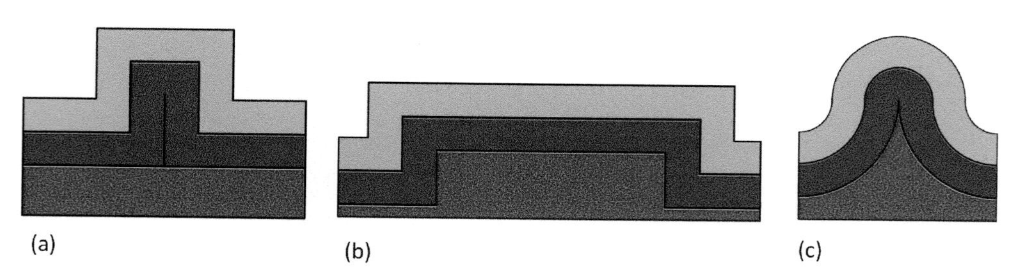

Figure 4.10 Diagram of box folds (a, b) and, for comparison, a concentric fold (c). All these are parallel folds and are illustrated here in their final state. Their kinematics, that is the change in them over time, is discussed in the text and detailed in Figure 4.12 for box folds and Figure 4.16 for the concentric fold.

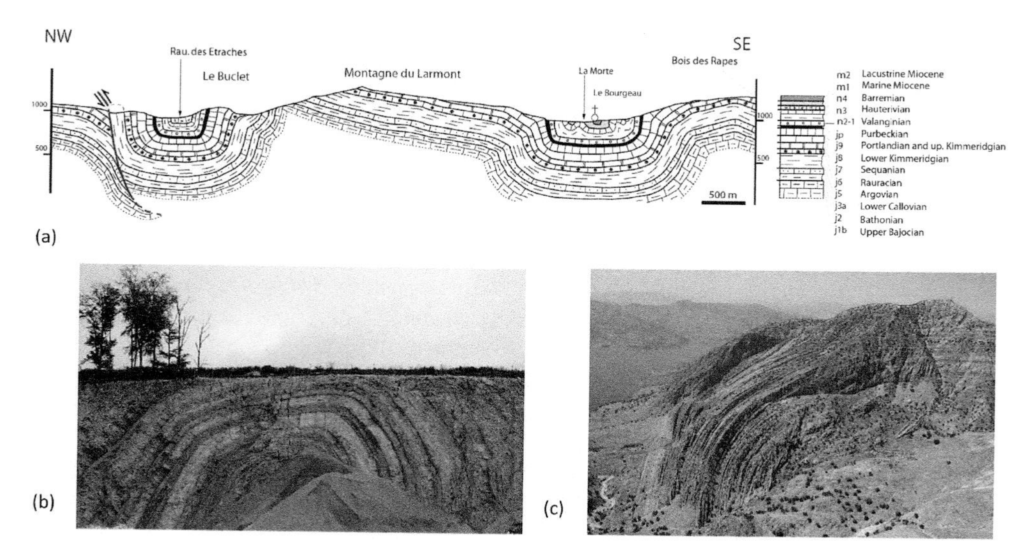

Figure 4.11 (a) Geological section through the Pontarlier map illustrating the box style of folding; (b) example of a box fold in Jurassic limestones of the Jura (Saint Ursanne area, Switzerland); (c) example of a box fold in Cretaceous limestones of the Izeh area (Central Zagros, Iran). It should be noted that in the examples illustrated the hinges are not kinks in the strict sense. To obtain a kink geometry the number of strata and, consequently, the possibilities for slip between the strata need to be multiplied (Fig. 4.4).

Source: (a) From D. Sorel and P. Vergely, *Initiation aux cartes et aux coupes géologiques* [Introduction to geological maps and sections], Dunod ed.; (b) photo © D. Frizon de Lamotte; (c) photo © D. Frizon de Lamotte

When a geological section includes box folds, the style most often applies on the scale of the section as a whole. This suggests that the vertical limbs develop immediately with this dip; otherwise one would expect to find intermediate stages with dips of between 0 and 90°. These limbs would not therefore be the result of progressive rotation but would gradually increase through transfer of material to the vertical limbs of the folds. This means that some hinges are not fixed with respect to strata. Material can therefore cross them during folding (Fig. 4.11).

Several scenarios for achieving such geometries can be imagined. Let us look at two of them that correspond to kinematics that can be easily modeled. In the first of these (Fig. 4.12a) the *kink bands* constituting the limbs are gradually and symmetrically supplied from the top of the structure. These limbs will grow until the top almost disappears when the two limbs encounter each other. The ductile core of the fold (usually salt) will therefore first fill and then completely empty in order to reach the final geometry. As a result the thickness of the basal detachment level will decrease and then increase. In the second scenario (Fig. 4.12b) material is delivered to the limbs laterally, and the length of the top of the anticline does not change. Thus, a box anticline will grow vertically and its ductile core will be fed by material coming from the detachment level, the thickness of which will therefore decrease. The limit in this case is exhaustion of the detachment level and/or the weight of the structure preventing further growth. We can obviously imagine intermediate mechanisms between these two extremes. It should be noted that in none of the cases is the relationship proposed by Goguel confirmed (Fig. 4.9).

Let us take things further with a geometric model that is still very simple but in which the dip of the fold limbs increases over time (Fig. 4.13). This is an anticlinal model consisting of isopach strata based on a detachment level that is able to flow. The model is constructed so that the excess surface area is triangular in shape. The apexes of this triangle are fixed hinges in the strata reference frame and therefore act like hinges. The base of the triangle is equal

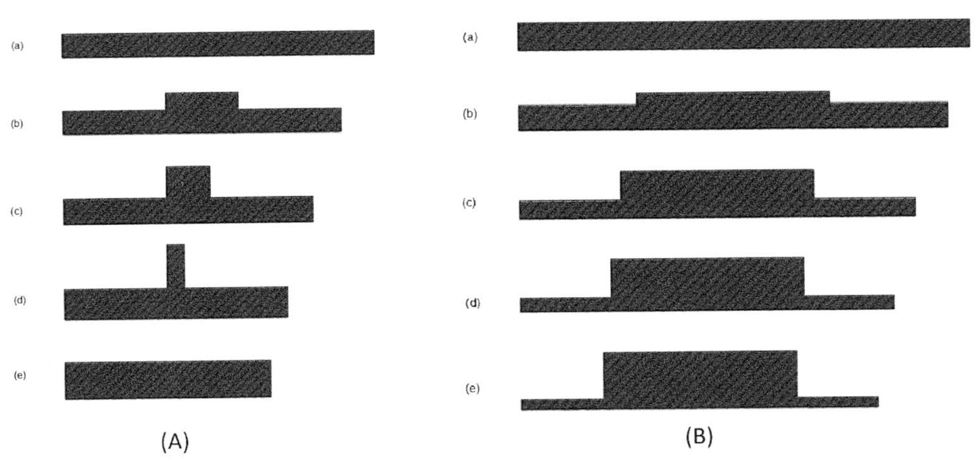

Figure 4.12 Diagrams explaining the change in shape and thickness of the detachment level when a box fold develops. In both columns (A) and (B) the sequence of sections [(a) to (e)] corresponds to serial sections in the 3D drawings used to print the models illustrated in Figures 4.15 and 4.16. In (A), which corresponds to the final state in the diagram of Figure 4.10a, the vertical limbs are fed from the top of the box fold. As a result, the fold rises and pinches quickly. The depth of the base relative to the bottom of the synclines will therefore decrease (between (a) and (b) and then increase from (c) to (d)). In (B), which corresponds to the final state in the diagram of Fig. 4.10b, the fold will on the contrary maintain the same width while gradually rising. The depth of the base in relation to the bottom of the synclines will gradually decrease from (a) to (e).

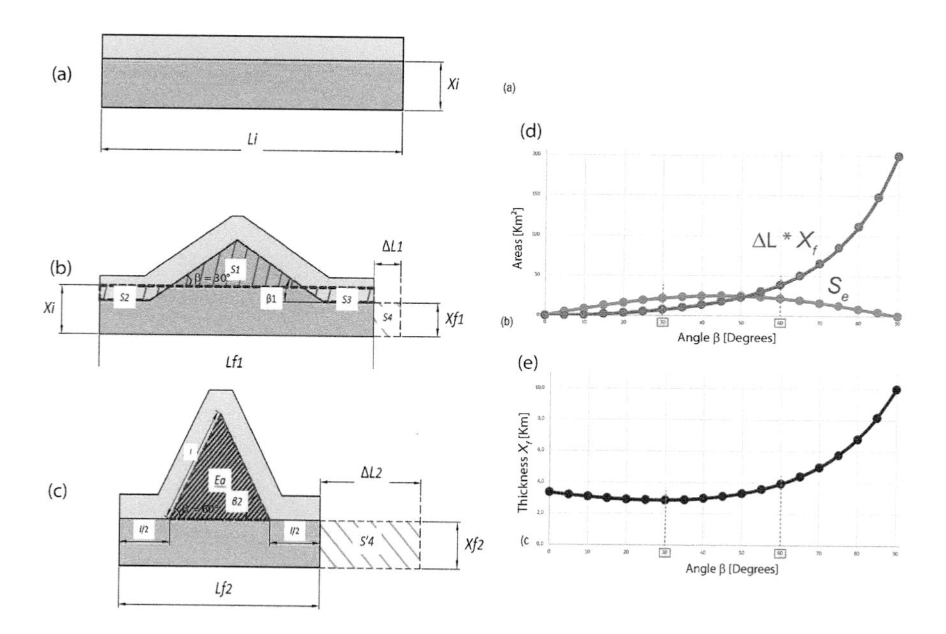

Figure 4.13 Kinematic model of an anticlinal fold with fixed hinges in the reference frame of strata. The curve shows us that the shortening calculated by measuring the lengths of the strata (ΔL) does not correspond (with a few exceptions) to that calculated by dividing the excess area by the depth of the detachment level (S_e/X). The successive stages in fold formation correspond to: (a) initial stage, (b) $\beta = 30°$ and (c) $\beta = 60°$; (d): comparison of surface areas *Ea* and $S4 = (\Delta L \times X_f)$ for each deformation increment; (e) change in X_f for each deformation increment.

to ΔL ($Li - Lf$). The other two sides (fold limbs) are made up of two segments of identical length (l). We can write (Fig. 4.10):

$$Ea = \tfrac{1}{2}\, l^2 (\sin\beta \times \cos\beta)$$
$$Ea = \tfrac{1}{4}\, (l^2 \times \sin 2\beta)$$

According to this model there is an increase in *Ea* (excess surface area) when the dip β of the fold limbs increases to $\pi/4$ and then a decrease in *Ea* beyond $\pi/4$.

Let us now look at how the surface area ($\Delta L \times X$) evolves over time. L (t) varies from Li to Lf, X (t) changes from Xi to Xf. Initially the surface area of the ductile layer (detachment level) is a rectangle of dimensions ($Xi \times Li$). For balancing purposes, this surface area must be maintained at every step up to the final state (see the box on balanced cross-sections). Creation of the structural relief associated with the anticline is therefore accompanied by subsidence of the two adjacent half-synclines. In other words, the base of competent strata moves toward the top of the basement under the synclines. On the basis of this observation, we can define several new surface areas in relation to the benchmark comprising the initial top of the salt layer: *S*1 is the surface area limited at the base by our benchmark and at the top by the top of the salt; *S*2 and *S*3 are the surface areas bounded by our benchmark at the top and by the top of the salt at the bottom. By construction, we have: $S2 = S3$ and $S1 = (S2 + S3)$, meaning that the salt which has accumulated in the core of the anticline is partly the result of transfer from the base of the synclines. If we compare surface area $S1$ with Ea, as defined by Goguel, an additional area $S4 = (Se - S1)$ appears. This surface area $S4$ therefore corresponds to the surface area that must be added to balance out the surface area of the detachment level. It is, by definition, equal to $(\Delta L.X)$ at every stage in the structure's development. ΔL is measurable. What we are looking for is to determine how X changes in relation to β.

We can say that at every moment of the development of the fold we have:

$$Ea + (X \cdot L) = (Xi \cdot Li)$$

$$X \cdot L = [(Xi \cdot Li) - Ea]$$

$$X = [(Xi \cdot Li) - (1^2 \cdot \mathrm{Sin}\ 2\beta/4)] / \mathrm{L}$$

This model enables us to suggest a kinematics and think about the 3D evolution of a fold. However, it is too simple because it produces an anticline with a geometry that is rare for an actual geological structure.

We will now consider the case of concentric folding above a low-friction detachment level (e.g., salt) (Fig. 4.10c). The final (observed) structure is a concentric fold with a cusp at the interface between the detachment level and the concentrically folded strata. Can we imagine the kinematics from this final state alone?

This is what we suggest in Figure 4.14. It is apparent that it is not possible to create the cusp observed between the detachment level and the first isopach layer in the final state

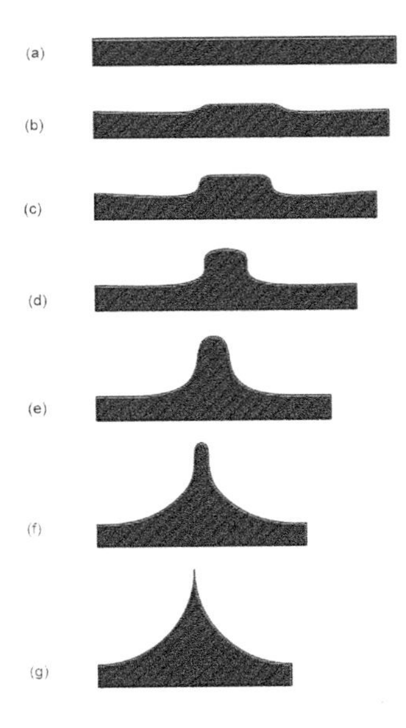

Figure 4.14 Diagrams explaining the change in shape and thickness of the detachment level when a concentric fold is formed. In its final state, the fold corresponds to the diagram in Figure 4.10c. It should be noted that the cusp only appears in the final stage of deformation. The sequence of sections [(a) to (g)] corresponds to serial sections in the 3D drawing that made it possible to print the models illustrated in Figure 4.17.

Figure 4.15 Photos showing pinching of the detachment level in the core of the anticlines until it forms a cusp. The two photos are helicopter views taken (a) in the Dezful Embayment and (b) in the Izeh area in the center of the Zagros range (Iran).

Source: Photos by © D. Frizon de Lamotte

(Fig. 4.10c and Fig. 4.14e) from the outset because this would lead to geometrical impossibilities (overlapping strata). It is therefore necessary to develop a central anticline with a rounded hinge and two adjacent half-synclines at the start (Fig. 4.14b). Creation of the structural relief associated with the anticlinal hinge is therefore accompanied by collapse of the two adjacent synclinal hinges, as in the previous case. The anticline will be amplified by rotation of its limbs and migration of the two adjacent synclinal hinges. At the interface between the detachment level and the concentrically folding strata, the hinge will pinch more and more until it changes convexity and generates a cusp (Fig. 4.15). The kinematics we suggest is certainly not unique. For example, the initial position of the lateral synclinal hinges can be adjusted. Bearing in mind that the cross-section is balanced, since surface areas and lengths are preserved at each step, it can be said that it is a geologically credible restoration.

In any event, a particular stage is reached when the cusp is formed (Fig. 4.14e, Fig. 4.15). Deformation could continue by gradually raising this point. Nevertheless, in general, the structure will evolve through the development of conjugate reverse faults on either side of the anticlinal hinge. The result will be an extruded core anticline evoking the "pop-up" structures mentioned at the end of Chapter 3 (Fig. 3.17), but this time without the need to involve the basement in the deformation.

4.4 3D models of detachment folds

We illustrate three 3D models of detachment folds. The aim is to illustrate the kinematic issues developed earlier (Figs. 4.12 to 4.14). These 3D blocks are balanced because we have checked that surface areas are preserved in each of the possible sections between the undeformed ("initial") states and the deformed ("final") states that constitute the parallel sides of the blocks. The other two sides are convergent, reflecting the increase in shortening between the "initial" and "final" states. Under these conditions, being able to build a harmonious block (without discontinuities) is in itself a test of balancing.

The model illustrated in Figure 4.16 is a box fold, which adopts the kinematics proposed in Figure 4.12a (vertical limbs fed from the top of the structure). Only the lower layer behaves as if it had low viscosity (concept of detachment level). The other two layers are isopachous. The lateral view shows the change in basal level thickness (decrease and then growth).

The model illustrated in Figure 4.17 is a box fold adopting the kinematics proposed in Figure 4.12b (constant width top and limbs fed laterally from adjacent synclines).

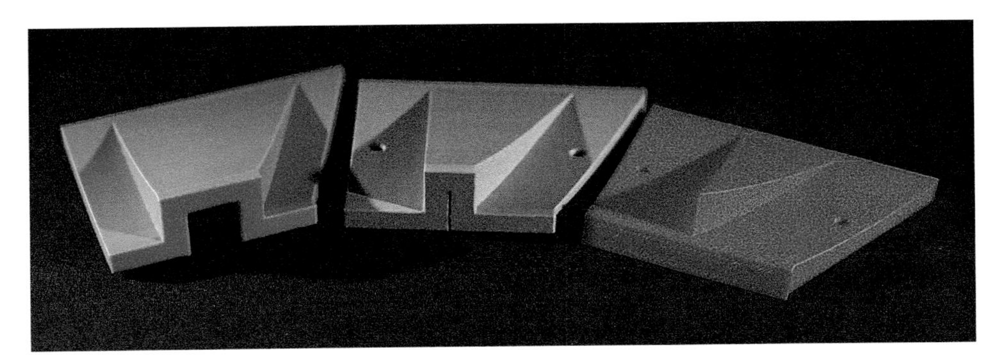

Figure 4.16 A 3D model of an isopach box fold such that the vertical sides are fed from the top of the anticlinal structure. The third dimension shows the evolution of the structure over time, that is, its kinematics (see Fig. 4.12a). Above, the assembled model; below, the three parts making up the model.

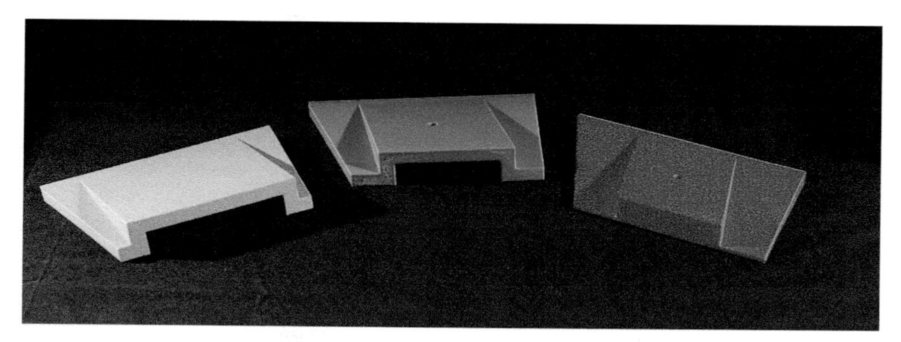

Figure 4.17 A 3D model of an isopach box fold such that the vertical limbs are fed laterally from the adjacent synclines. The top maintains its width and is raised over time. The third dimension illustrates the kinematics (Fig. 4.12b). Above, the assembled model; below, the three parts making up the model.

The side view shows the progressive decrease in the thickness of the detachment level.

The model shown in Figure 4.18 is an isopach cylindrical fold in accordance with the kinematics proposed in Figure 4.13. The view of the ductile basal level shows the evolution of the central anticlinal hinge over time and in particular the change in the concavity of the flanks just before the final stage. Unlike the previous models (Figs. 4.16 and 4.17), these kinematics are not unique, and the position of the two synclinal hinges adjacent to the central anticline at the beginning of the folding could be varied.

Figure 4.18 A 3D model of a cylindrical fold. Note the evolution of the style at the interface between the red and blue strata. The starting point is a boxlike fold evolving toward a rounded and then acute hinged fold. The third dimension illustrates the kinematics (Fig. 4.14).

4.5 The role of the basement: "thin-skinned" tectonics and "thick-skinned" tectonics

The study of detachment folds poses the important question of the role of the basement during compressive deformation and more generally in orogenesis. In all the previous examples, the basement was hardly mentioned or rather it was considered to be unaffected by the deformation and illustrated as a rigid support. This is referred to as cover tectonics or, better still, as thin-skinned tectonics, to emphasize that only the "skin," the surface layer that forms the cover, is involved in the deformation.

The Jura section in Figure 4.7 illustrates this concept of thin-skinned tectonics magnificently. It shows a set of folds completely disconnected from its basement over a distance of about 40 km, thanks to a major detachment level located in the evaporites of the Triassic. The basement structures are much earlier, dating from the late Paleozoic, the Variscan orogenesis. It is in fact the basement of Western Europe, which consolidated around 300 Ma.

If, mentally, you try to unfold the section, that is, to return the sedimentary strata to their initial horizontal position, there is immediately a problem. There is too much cover in relation to the basement! Such an initial configuration is of course impossible. It is therefore necessary to accept that laterally (toward the southeast), deformation of the basement compensates for the deformation deficit of the basement under the Jura. Thus, all sections through the Alps including the Jura show an outer thin-skinned part (the Jura) supplemented

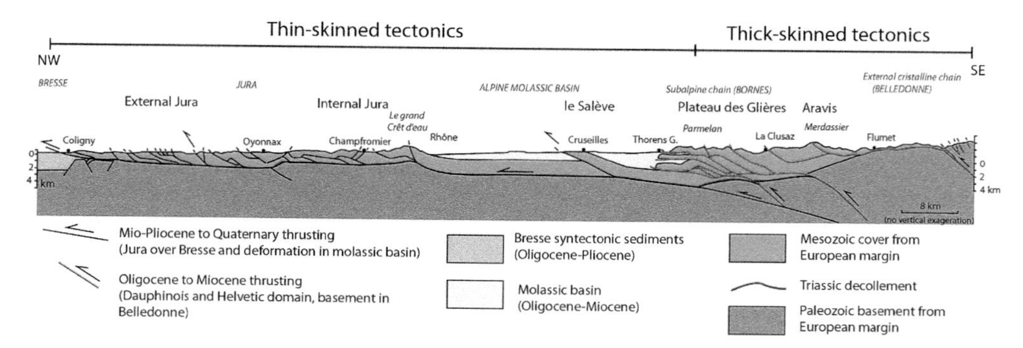

Figure 4.19 General section through the Jura and the outer Western Alps illustrating the concepts of thin-skinned tectonics (Jura) and thick-skinned tectonics (outer Alps) and the necessary connection between these two deformation modes.

Source: Modified after M. Renard, Y. Lagabrielle, M. de Rafélis and E. Martin, *Eléments de Géologie* [Elements of Geology], Dunod ed.

by an inner "thick" part (the outer Alps), where the basement is clearly involved in the deformation (Fig. 4.19). In the literature, this is referred to as thick-skinned tectonics. It indicates that the "skin" involved in the deformation is thick because it includes not only the sedimentary cover but also the underlying basement, or at least part of it. Involving the basement in deformation means that the detachment surface at the base of the cover must be connected to deeper detachment in the basement. Such a connection needs a fault to be present between the two detachment levels.

The reasoning we have just made on the need to compensate for the deformation deficit of the basement under the Jura in the Alps lies at the root of the techniques for balancing geological sections (see the box on this topic). These techniques that make it possible to validate a section, to consider it likely even without being certain of its accuracy, will not be developed in detail in this book, but it is important to know their principles. It should be noted that a section through a mountain range, and not just a portion of it, cannot be of thin-skinned style over its entire length. This style can only apply to a portion of the whole. The Jura is therefore an element of the Alpine chain. It can only be understood in its entirety by including the Alps (Fig. 4.19).

When we try to build a complete section through a mountain range like the Alps, and it is an exciting exercise, we come up against the question of how deeply we should pursue the thrusting responsible for its surface structure. We have seen that the lower crust has ductile behavior and can therefore act as a detachment level. But beyond that? The fact that the lower crust and even the upper mantle may be carried to the surface by thrusting in some chains indicates the possible existence of very deep detachment in the sequence of envelopes constituting the tectonic pile. Despite this, there must be a lower limit. This detachment level able to create a lithospheric root independent of the crustal root must not be very far from the Moho.

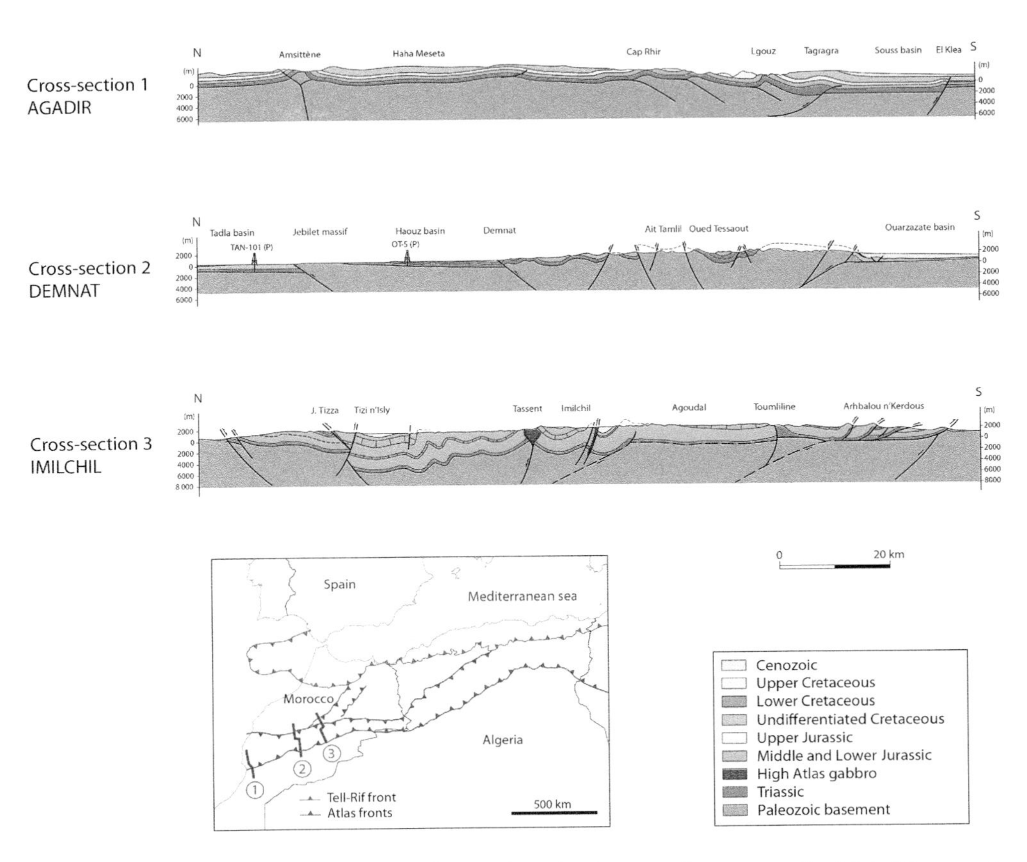

Figure 4.20 An example of a thick-skinned tectonic style, three sections across the High Atlas mountains (Morocco) showing that the basement is everywhere involved in deformation.

Source: The sections have been redrawn after Teixell *et al.* (2003).

If a mountain range cannot be entirely thin-skinned, it can, on the other hand, be of thick-skinned style over its entire width. This is often the case with intracontinental chains resulting from the tectonic inversion of an aborted rift whose two edges serve as a buffer during compression. An example of this tectonic style is the High Atlas mountains in Morocco (Fig. 4.20), where, despite an effective detachment level, the basement is involved in deformation near the two mountain fronts and forms the highest peaks.

Chapter 5

When folds and faults interact

Fault-related folds and fold-and-thrust belts

5.1 The concept of fault-bend folding

The term fault-bend fold was coined by John Suppe (1983), but the concept is much older. This model was originally proposed by John Rich (1934) to describe a regional fold in Powell Valley in the Appalachians (US) (Fig. 5.1). The idea is that the geological strata are bent at the top of a "staircase" thrust, presenting a succession of segments parallel to the strata (décollements called flats) connected by oblique segments (faults called ramps). Thus, a thrust gradually climbs through a stratigraphic sequence generating ramp anticlines, the size of which depends primarily on the length of the ramp. Folding is not a result of buckling, as indicated previously for detachment folds (Fig. 3.1), but from a gradual bending of the strata to match the stepped shape of the thrust over which they move. Subsequently, this stair-step geometry of fault-bend folds was recognized in numerous examples in the field (Fig. 5.2) or in the subsurface, thanks to industrial seismics (Fig. 5.3).

Going back to the possible mechanisms, let us imagine a very simple model of a package of strata approaching a ramp (Fig. 5.4a). On reaching the ramp, the deformation is accommodated by the development of a kink. It is to be hoped that the model will maintain the thicknesses and lengths of the strata. For it to do so, the kink plane has to bisect the angle between the ramp and the flat. Deformation will occur whenever the thrusting upper strata have to adapt to a flat-ramp or ramp-flat transition. At the foot of the ramp, the strata will rise up to adapt to the shape of the thrust fault (Fig. 5.4b). This imposed bending is accompanied by a deformation that corresponds to a flexural slip. In this way, the strata will fold at the flat-ramp transition before progressing onto the ramp. As in Chapter 4, we may postulate a certain hinge migration in the reference frame of the strata. Conversely, the hinge remains fixed in the reference frame of the autochthon, that is, in this case, to the static lower flat-ramp transition.

Once they reach the top of the ramp, the strata have to fold back to accommodate this new transition and continue along their path on the upper flat. This results in a new bend in the opposite direction to the previous one (Fig. 5.4c). The configuration is the same as before, with a mobile hinge in the reference frame of the strata but fixed with respect to the ramp-flat transition of the autochthon. Finally, it is as if the strata were crossing the hinges, that is, were folded and then unfolded when they reach the upper flat. Note that in the final stage this new deformation cancels out the previous one, recorded when the strata are engaged on the ramp.

In the same way as for décollement folds, the folding conditions at the top of the ramp will depend on the thickness of the strata and the effectiveness of the décollements. The lower the

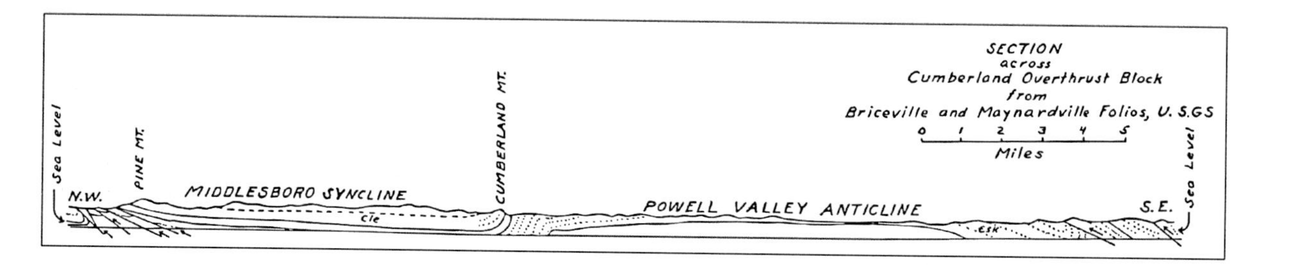

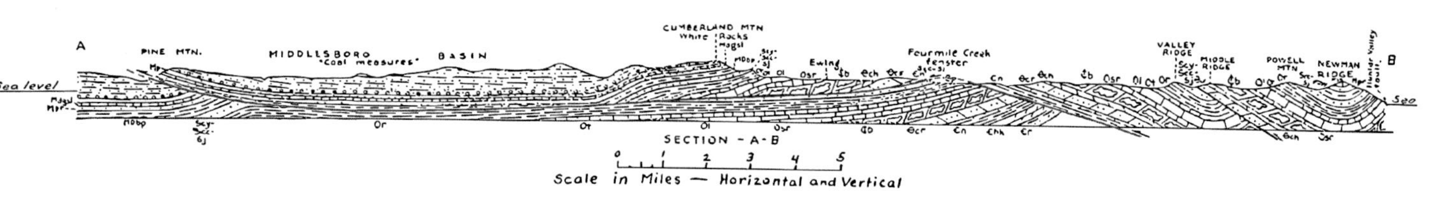

Figure 5.1 The "Powell Valley" anticline and the history of the concept of fault-bend folding. Above, Butts' early (1927, in Rich [1934]) interpretation showing a syncline and an anticline box fold wedged between fault zones. Rich's interpretation below is quite different. Detailed mapping enabled him to show the existence of a tectonic window at the heart of the Powell Valley anticline, from which he deduced its particular staircase thrust shape (highlighted with red dashes) and everything else ensuing from this.

Source: Both cross-sections are from Rich (1934)

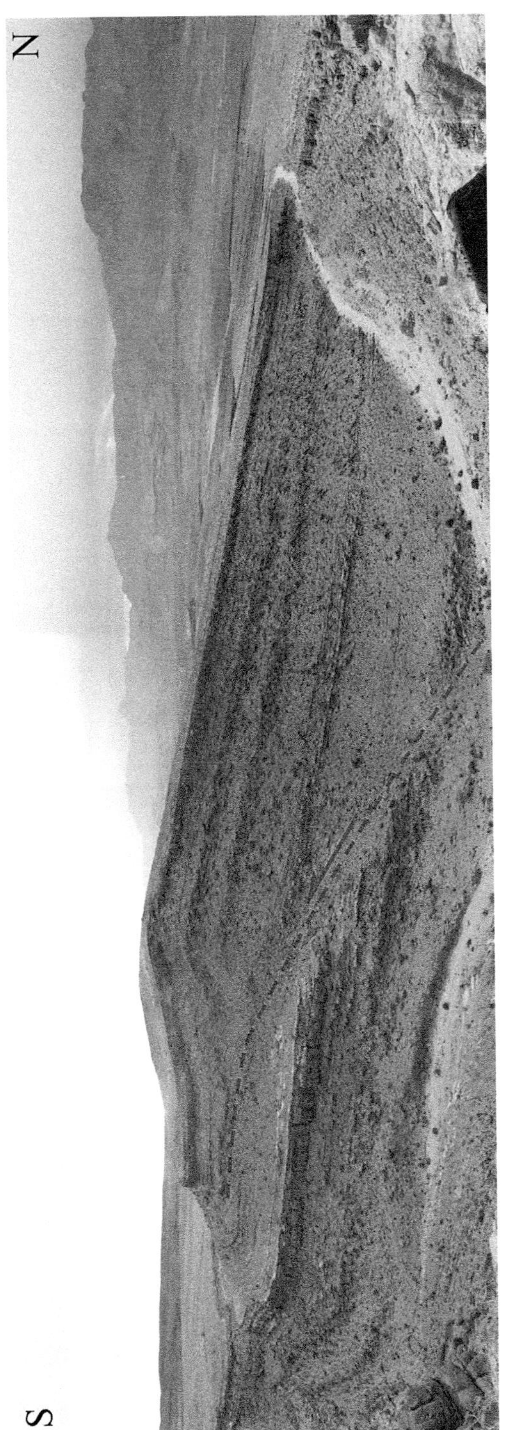

Figure 5.2 Example of a fault-bend fold in the Cretaceous terrains of the southern limb of the High Atlas (Goulmima, Morocco).

Source: Photo © Pascale Leturmy

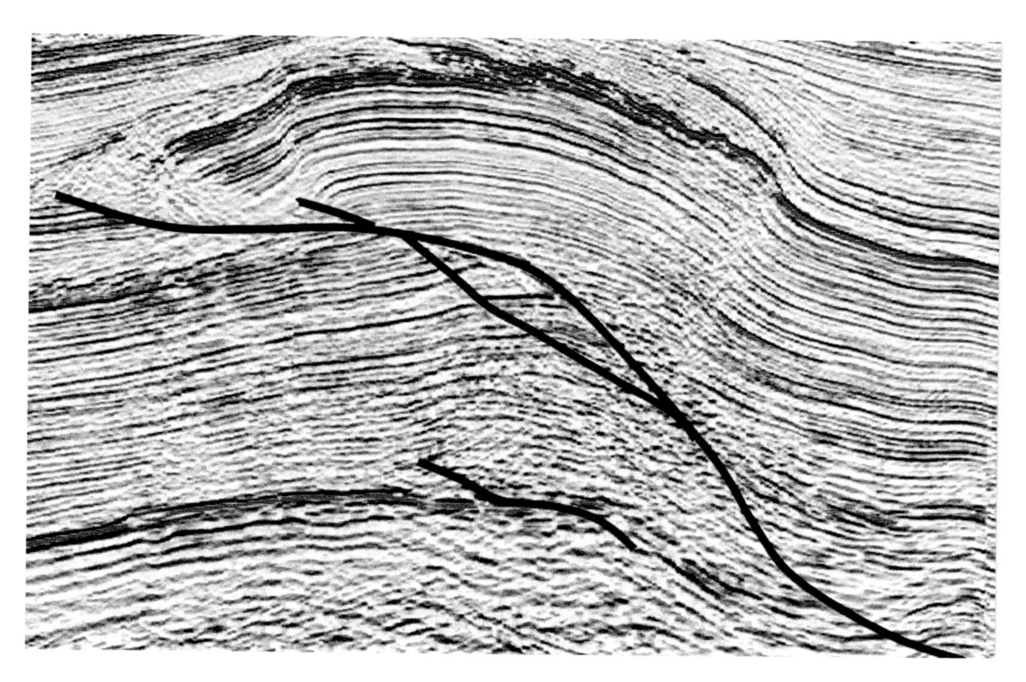

Figure 5.3 Seismic profile illustrating the concept of fault-bend folding.

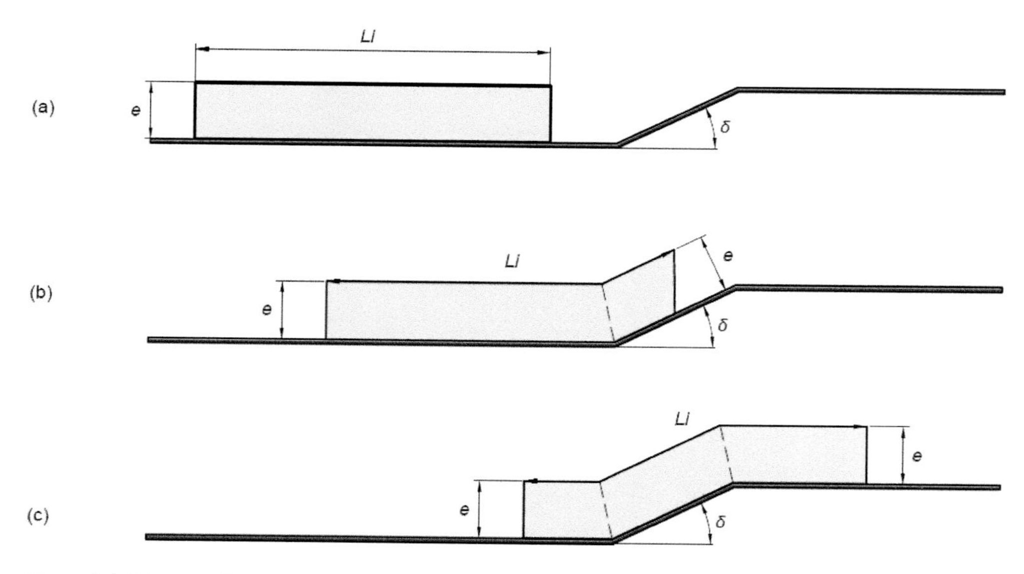

Figure 5.4 Diagram showing the deformations expected at each flat-ramp, then ramp-flat transition. Forward simple shear occurs at the lower flat-ramp transition, and then a backward simple shear at the upper ramp-flat transition. It can be seen that the two stages of deformation are superimposed and thus cancel each other out when the package of strata reaches the upper flat.

number of strata, the more dispersed the hinge will be. Conversely, the more strata there are, the more localized the hinge will be. Indeed, a large number of strata will lead to bed-on-bed slipping, thus facilitating the localization of the deformation. Ultimately, for an infinity of strata, the hinge will be a kink, such as the one shown in Figure 4.5. Localization is thus at its maximum: it is in the end concentrated on a single plane, the kink plane, which corresponds to the axial plane of the kink fold. It was precisely by using this kink method to simplify the problem that John Suppe (1983) formalized fault-bend fold geometry. As has just been seen, in order to preserve the thicknesses, all kink planes must be considered as bisecting the angles between the flats and the ramp. The frontal angle must also be such as to maintain the lengths of the strata. John Suppe established the following relationship between δ (ramp angle) and γ (upper angle of the frontal triangle), as shown in Figure 5.5:

$$\tan\delta = \sin\gamma / [2\cos^2(\gamma/2) + 1]$$

In the initial stage, the ramp intersects the strata at its hanging wall and at its footwall, defining cutoffs. The footwall cutoff of the ramp will remain stationary. Conversely, the hanging wall cutoff will move together with the strata (Fig. 5.6). The identification of hanging wall and footwall cutoffs is very important. Indeed, when recognized on both sides of a thrust, the displacement (slip) along this thrust can be measured. Conversely, it is safe to say that recognition of a cutoff in the footwall of a fault means that it can be called a ramp, even

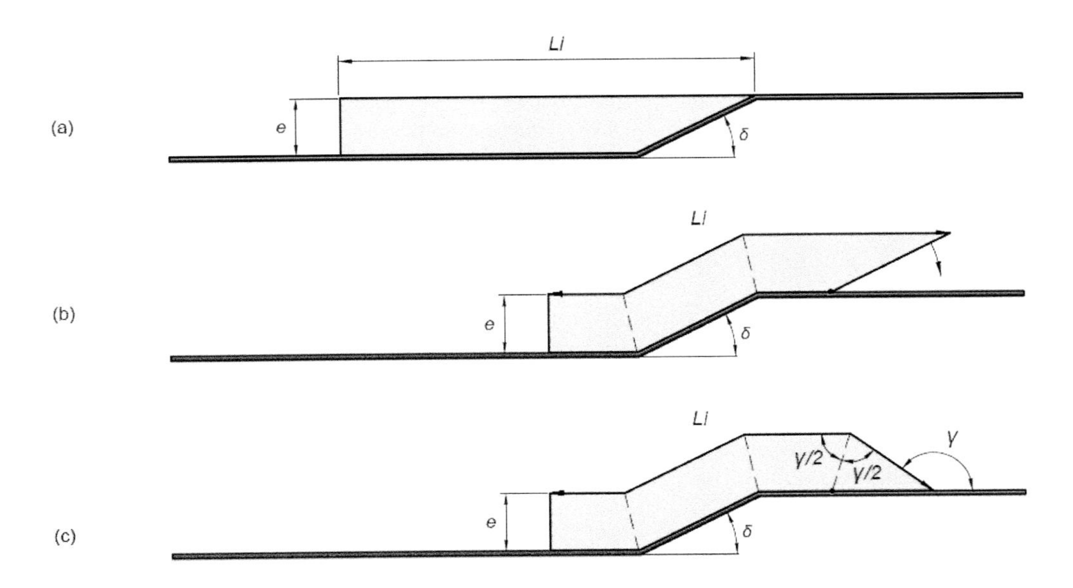

Figure 5.5 Diagram illustrating the geometry, which allows the frontal triangle to be folded back by means of the kink method while maintaining its lengths and surfaces. The graph shows the relationship between the angle of the ramp (δ) and the angle (2γ) representing the additional angle of dip of the front limb. There is no solution for a value of $\delta > 30°$. The angle of dip of the front limb is always $< 60°$. This mode of folding is therefore unsuitable for modeling folds whose front limb is very steep or reversed.

Source: Modified from Eric Mercier, 1996

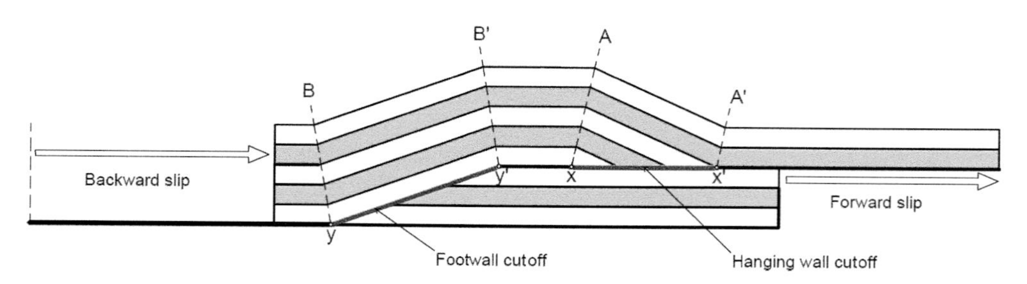

Figure 5.6 Model of fault-bend folding using the kink method, illustrating the concept of the concept of hanging wall and footwall cutoffs, classification of hinges and the backward and forward slip of the fold. In the case shown here, which corresponds to stage (d) of Figure 5.7, 87% of the backward slip is transmitted forward on the upper flat; 13% of the backward slip is thus accommodated in the fold itself.

if it has lost its initial attitude, that is to say it has been folded or simply tilted at a later stage. This is the case on the southernmost fold on Buxtorf's Jura cross-section (Fig. 4.7): the ramp has become very steep (even vertical) but retains a footwall cutoff, showing that its current attitude is the result of later folding allowing its tilting.

5.2 Kinematics of fault-bend folding

To trace the kinematics of fault-bend folding, we will focus on describing the evolution of various hinges over time (Fig. 5.7). These include a stepped thrust fault (flat-ramp-flat), an autochthonous domain (under the thrust) and an allochthonous domain (at the hanging wall of the thrust). The shape of the thrust fault requires four hinges (B, B', A, A') (Fig. 5.7) whose history can be described. Hinge B is fixed at the lower flat-ramp junction. It is therefore fixed in the reference frame of the autochthon but mobile in the reference frame of the allochthon. In fact, the allochthonous strata are folded when crossing hinge B and then transported as a rigid body on the ramp. Hinge B' is, in turn, fixed in the reference frame of the allochthon but mobile in the reference frame of the autochthon. At first, it is combined with B but then detaches from it to settle on the upper flat-ramp transition, where it changes status and becomes fixed in the reference frame of the autochthon. Hinge A also has a two-step history. At first, it is fixed at the upper flat-ramp intersection and is then crossed by the strata pushed previously onto the ramp. When hinge B' reaches the top of the ramp, hinge A changes role and becomes fixed in the reference frame of the moving strata on the thrust fault. As for hinge A', which delimits the frontal panel, it remains constantly fixed with respect to the allochthon (it is never crossed by the strata) and therefore mobile with respect to the autochthon.

It should be noted that some hinges in this model (B' and A) have complex histories, with a change in status during their evolution (Fig. 5.7). All the material constituting the folded part of the allochthon has, at a certain moment in its history, crossed a hinge except for a triangle limited by hinges B' and A at the moment when B' reaches the top of the ramp. This triangle is translated rigidly, with a change in trajectory at the ramp-upper flat transition. This is also the time when the frontal panel stops growing to begin its translation onto the upper

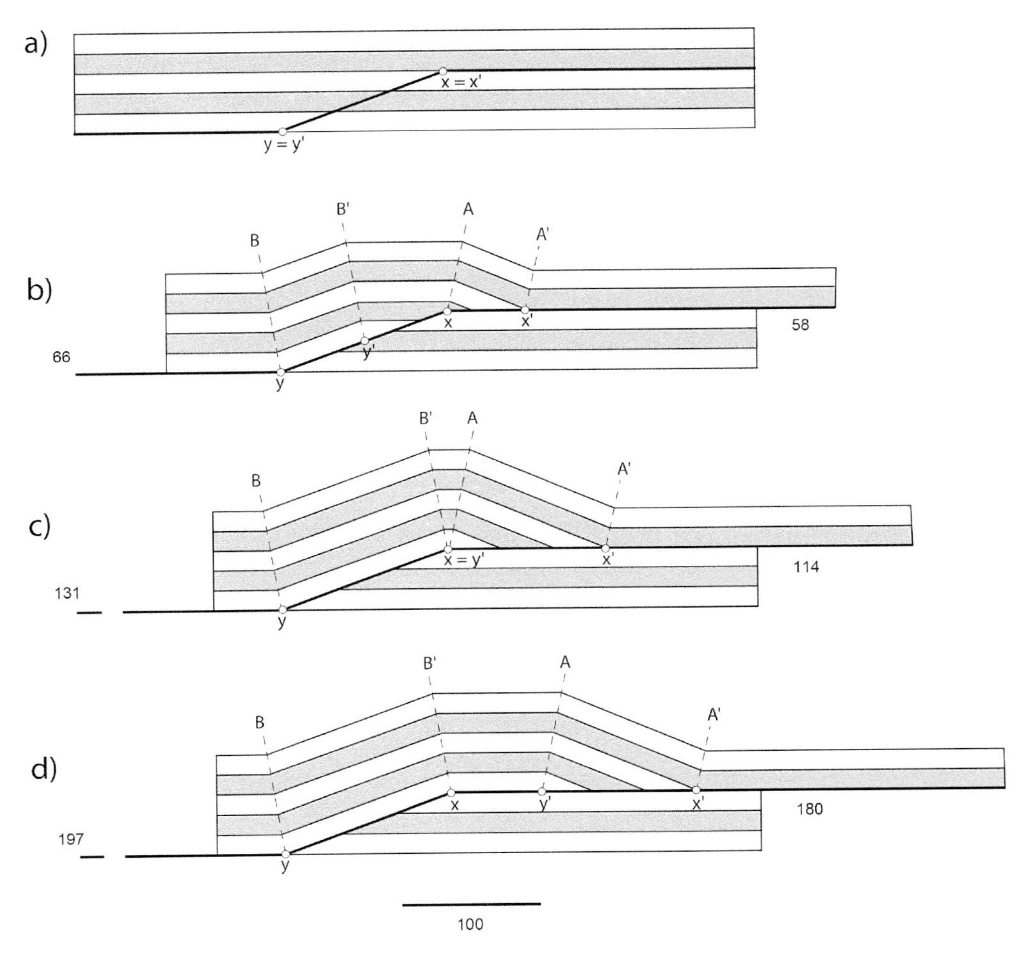

Figure 5.7 Kinematic model of a fault-bend fold using the kink method (Suppe, 1983). The numbers indicate the slip imposed at the rear of the structure and the forward transmitted slip, respectively. See explanations in the text. Cross-sections (a) to (d) correspond to the kinematic steps and to some cross-sections in the 3D model of Figure 5.13.

Source: The drawing is modified after Diégo Buil (2002)

flat and when the fold reaches its full height. From then on, its growth is limited to a widening of the fold central panel (roof). If erosion progressively attacks the superstructure of the system, thrusting may continue for a long time. Conversely, if erosion is weak, the weight of the allochthon will soon inhibit thrust propagation, which will be relieved by the development of a new, more frontal thrust fault. This type of sequential thrusting is notably imaged by analog experiments in "sandboxes" (see the box on analog models).

The model assumes that the thrust predates any other deformation, or, more precisely, that the propagation of the thrust is instantaneous throughout its length at the start of the deformation. In other words (see Chapter 3), $P/S = \infty$ at the initial stage of development of the anticline, then P/S = 0 during the development of the fold. Such behavior is obviously unrealistic: it is a model. In reality, the flats correspond to areas of weakness inherited from the sedimentary history (the concept of décollement level developed in Chapter 4). They do not therefore need to be created but simply activated. Conversely, unlike in the case of décollement folds (Chapter 4), they do not need to be thick but simply to exist. As regards the ramps, they are short segments connecting two flats. It can be presumed that they propagate very quickly.

The slip applied at the back of the fold is partially accommodated in the fold, but another part is transmitted forward on the upper flat (Figs. 5.6 and 5.7). As a result, unless the upper flat is the topographic surface, the development of a new, more frontal structure is necessary to accommodate this transmitted residual slip, then possibly another and so on until the topographic surface is reached. Except in very rare cases, a thrust fault will not descend into the stratigraphic pile. Such a configuration is most often the result of the existence of a previous structure, a fold on which the thrust abuts and penetrates. The thrust is then known as "out of sequence" with respect to this previous structure.

5.3 Fault-propagation folding: geometry and kinematics

The idea of a fault-propagation fold is rooted in the observation of the existence of faults that die out in a fold. This observation can be seen in maps and in cross-sections. Gallup's cross-section of the Turner valley in Alberta (1951) shows a thrust (a ramp) at the heart of a fold with a short and very upright front limb (Fig. 5.8). By identifying the different strata on either side of the fault, it can be seen that the higher up you go in the structure, in this case the stratigraphic pile, the lower the difference between strata of the same age. In other words, the displacement gradient is between the tip of the ramp, where the displacement is zero, and the foot of the ramp, where it is at its maximum value.

On a map, thrust-faults often have a bowed trace and die out laterally within folds. By measuring a great number of thrust faults, the Canadian geologist David Elliott established

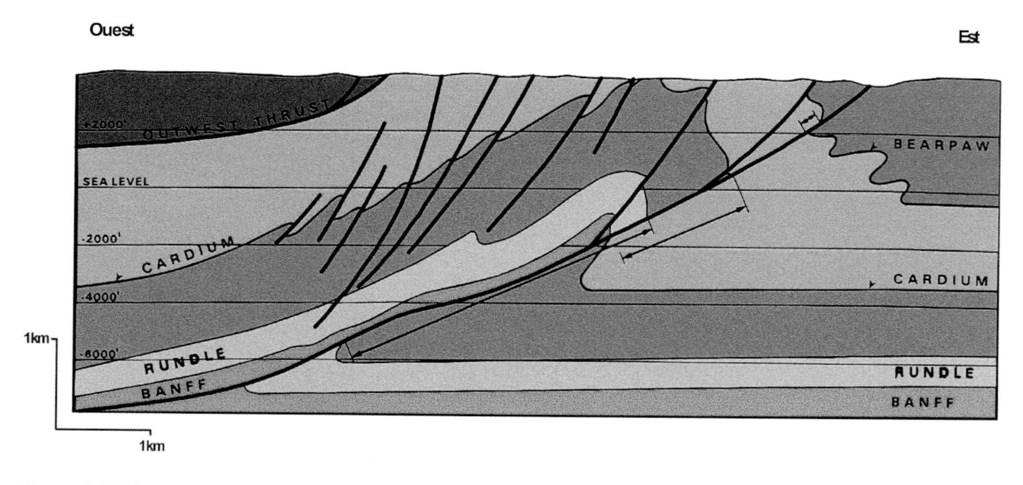

Figure 5.8 Historical cross-section of the Turner valley (Alberta, Canada) illustrating the concept of fault-propagation folding. The names are those of the geological formations of the Canadian Rocky mountains.

Source: After Gallup (1951, modified)

an empirical rule, the bow-and-arrow rule, according to which the ratio between the arrow and the string of a curved thrust fault was arofgund 10% all over, regardless of the scale of observation. This has major implications. The bow-and-arrow rule indicates, first, that the lateral propagation of a thrust is 10 times faster than its frontal propagation. It also indicates that at each stage of its lateral propagation, the thrust intersects the fold that has developed at its termination. Thus, the two lateral folds located at the terminations of the thrust fault were fossilized at the end of its progression and are indicative of the history of the thrust fault as a whole.

It is easy to model the formation of a fault-propagation fold by means of the kink method (Fig. 5.9). When the fault begins propagating by intersecting the strata from the lower flat, two kink-bands are formed almost instantaneously and thus determine four panels separated by kink planes defining four hinges (B, B', A, A'). Hinge B, similar to the one modeled for the fault-bend fold (Fig. 5.7), is anchored on the flat-ramp intersection. The strata cross the kink plane and undergo rotation, keeping the strata parallel to the ramp. Axial plane A', which limits the front panel of the fold downwards, is fixed at the tip of the ramp but, generally speaking, is not parallel to it. This frontal hinge is also traversed by strata initially located at the front of the structure and incorporated in the frontal panel of the fold. This hinge, together with the ramp, thus represents the limit of the autochthonous domain at a given moment. The mechanism involved at the forelimb of the fold can be illustrated by the image of a carpenter's plane, the blade (ramp) of which cuts a piece of wood and removes a chip (the forelimb of the fold).

Hinges B and A' are both traversed by the moving strata. Hinge B is fixed in the reference frame of the autochthon, while hinge A' is mobile in this benchmark because autochthonous material can be incorporated into the structure at any moment. The transfer of material from the autochthon to the allochthon is a major difference with the fault-bend fold. Hinge B' limits the backward panel of the fold upward. It is crossed by strata coming from the upper triangle, whose surface is reduced over time and, at the same time, pushed upward by the strata having crossed hinge B. This hinge is therefore doubly mobile in relation to both the autochthon and the allochthon. Hinge A limits the frontal panel upward. It is fixed (in reality almost fixed, but this is a minor detail) in the benchmark of the strata. Together with the horizontal hanging wall of the fold, hinges B' and A form an upper triangle whose surface decreases over time due to the migration of hinge B'.

This model presents several geometrical features that help in its construction. The two hinges B' and A converge at a point which, at each stage of development, is on the stratum which is reached by the ramp at the same time. One of the consequences is that the upper triangle disappears when the ramp reaches the surface (Fig. 5.9f). The other consequence is that there is a unique relationship between the surface geometry and the deep geometry. This feature also makes it differ significantly from the fault-bend fold. In the latter, in fact, the position of the lower and upper flats cannot be determined by the simple surface geometry, nor can the value of the slip transmitted at the front of the structure.

It should also be noted that the propagation of the ramp, or its length, is twice the slip applied at the back of the structure. The model thus sets a P/S ratio = 2 (see Chapter 3 for the definition of the P/S ratio). This is due to the fact that half the strata engaged on the ramp were originally located on the lower flat and half are strata deriving from the gradual erasure of the upper triangle. As a consequence, in terms of relief creation, fault-propagation folding (Fig. 5.9) is more efficient than fault-bend folding (Fig. 5.7). Note that this is not necessarily an advantage, since the gravitational forces related to the weight of the structure tend to inhibit its development.

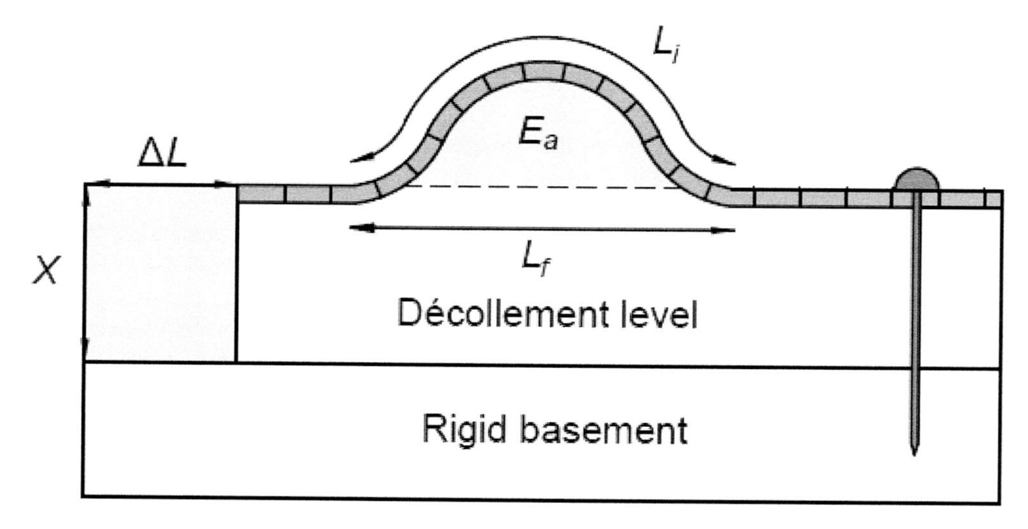

Figure 5.9 Kinematics of the fault-propagation fold according to the kink method (Suppe, 1985). The drawing is modified after Diégo Buil (2002). The numbers indicate the slip imposed at the rear of the structure. All this slip is accommodated in the fold (compare with Fig. 5.7). Cross-sections (a) to (f) correspond to the kinematic steps and to some cross-sections in the 3D model in Figure 5.14. Note the inverted forelimb, a geometry that is impossible to obtain for a fault-bend fold (Fig. 5.5).

5.4 Variations and limits of kinematic models of fault-bend folds

A model is by its very nature an idealized and therefore simplified representation of reality. Its aim, in our opinion, is not to take into account the subtleties of detail of a real structure but to shed light on the mechanisms involved, to force us to think. The initial kinematic models for fault-bend folds (Figs. 5.7 and 5.9) have given rise to many objections with regard to their simplified or even simplistic nature.

There are three main criticisms regarding kinematic models using the "kink method." The first concerns the realism of the shape of the folds obtained in this way. The second relates to the decision to keep the thickness of the strata. The third is the P/S ratio, which, as we have seen, is imposed by the models. The first issue has already been answered by showing, on the one hand, that "kink" folds did actually exist in nature in highly layered environments (Fig. 4.4), and on the other hand that the shape of the fold could be modified without challenging the fundamentals of the model: this is the case, for example, of box folds whose hinges are rounded (Fig. 4.11). The second and third issues are more problematic. In fact, the observation of natural objects frequently shows a thickening of the limbs of the folds associated with the propagation of a fault (Fig. 5.10). Conversely, it is clear that there is no reason to consider that the P/S ratio set by the model corresponds to a reality.

Many attempts have been made to "refine" the models and make them more realistic in order to address such objections. These either proposed secondary evolutions to enable

(a)

(b)

Figure 5.10 Two examples of fault-propagation folds showing a thickening of the front limb of the fold. (a) Dead Sea site; note also the development of a footwall syncline under the ramp; (b) Lagrasse (Corbières) site, a pocket fold located under the main flat of the famous Lagrasse fold.

Source: (a) Photo © Jean-Claude Ringenbach; (b) photo © Dominique Frizon de Lamotte

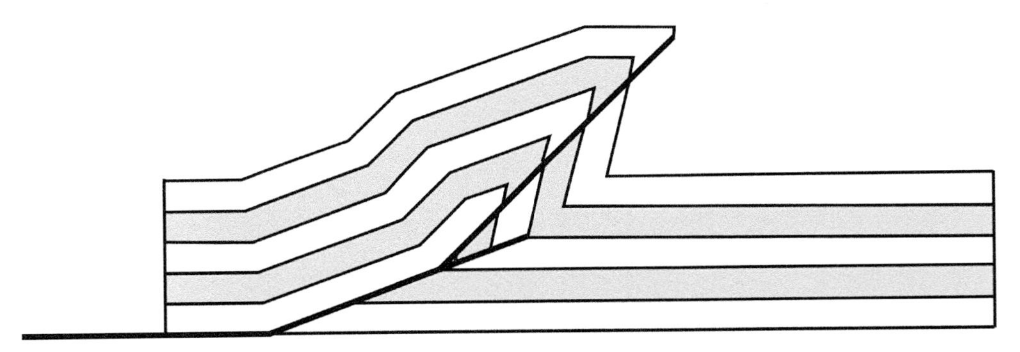

Figure 5.11 Fracture of the front limb of a fault-propagation fold. The fault that breaks the upright forelimb is supposed to propagate instantly (very strong P/S ratio). The model therefore adds a fault-bend folding mode associated with the development of the fault in the initial propagation mode.

Source: Model originally developed by Eric Mercier (1996)

the margins of the initial model to be modified, or else they relaxed certain constraints by allowing, for example, the thickening or thinning of the strata. The category of secondary evolutions includes the classic case of the fracture of the front limb of a fault-propagation fold (Fig. 5.11). This model takes into account the frequent observation of the development of late faults intersecting front limbs with steeply dipping folds.

A trishear kinematic model can be used to vary the strata thickness and the P/S ratio. Figure 5.12 shows a model designed in the laboratory by one of our doctoral students, Diégo Buil. The idea, generalized after other authors, is that the shear related to the crossing of a hinge is not accommodated regularly along the kink plane but in a fixed triangle zone either at the flat-ramp transition or at the end of the moving ramp. The front triangle zone is mobile, with the end of the ramp just like the "ductile bead" discussed in Chapter 3 (see Figs. 3.12 and 3.13). The main difference is that the ramp intersects the strata obliquely. The triangle zone is in fact an area that allows the strata of the autochthon to be transferred toward the allochthon; the autochthon is fixed, while the allochthon moves at a constant speed parallel to the ramp. The change in speed is accommodated along the ramp, going immediately from zero to a stationary speed typical of the allochthon. In the triangle zone, in contrast, there is a velocity gradient and the strata undergo deformation to gradually accommodate the transition from immobility to the velocity of the allochthon. In practice, the strata undergo rotation and thickening as long as the dip of the strata remains less than 90°; then, when the dip reaches the vertical and reverses, they undergo further rotation, but this time with thinning.

Deformation stops as soon as the material leaves the triangle zone. The deformed strata then either settle under the ramp or are transported on it as rigid bodies. Thus, the anticlinal and synclinal folds observed on either side of the ramp are not drag structures related to the progression of the ramp but structures that originate at the front of the ramp (in the triangle zone) and are then intersected by it.

The status of the rear triangle zone is different, in that it is fixed in the benchmark of the autochthon. Conversely, it is also a transfer zone allowing the strata to pass from the lower flat to the ramp by changing both speed and trajectory. As has been seen, the speed of movement on the ramp is stationary and parallel to the ramp. At the entrance of the triangle zone the strata have the same velocity ($V0$) and will all have equal velocity ($V1$) at the exit. To accommodate this change in velocity while preserving the surfaces, each stratum must both rotate and change thickness. The change in thickness of the strata depends on the inclination of the triangle zone. To obtain thickening, the angle bisector needs to be tilted backwards, while thinning is obtained if it is tilted in the other direction. In the first case, $V1 < V0$; in the second, $V1 > V0$. The angle of the triangle zone will manage the time required for the passage from flat to ramp and thus, finally, the curvature of the hinge at the hanging wall of this transition. If the angle is low, the hinge will be sharp (for a zero angle, the hinge is a kink plane); if the angle is high, the hinge will, on the contrary, be wide and progressive.

The limitation of this type of "refined" model is that virtually anything can be obtained by adjusting the various parameters. Indeed, by changing the value of the P/S coefficient and the coefficient of the angles and tilting the triangle zones forwards and backwards, it will always be possible to make a model fit the structure to be described. This is known as creating an *ad hoc* model. In doing so, we believe the main interest of a model is lost: its predictive nature. That is why our team has always favored the most simple and robust kinematic models; specifically, those using the kink method or a derivative of this method. They enable complex geological cross-sections to be balanced quickly and efficiently, while refinements can come later and comprise ornamentations whose possible geometries, or "structural style," are proposed by the geologist (see on this topic the paper by Rob Butler and colleagues [2018]). Indeed, inaccessible or hidden complexities can often be imagined by analogy with known structures. Making analog models can thus be of considerable help in proposing elegant solutions.

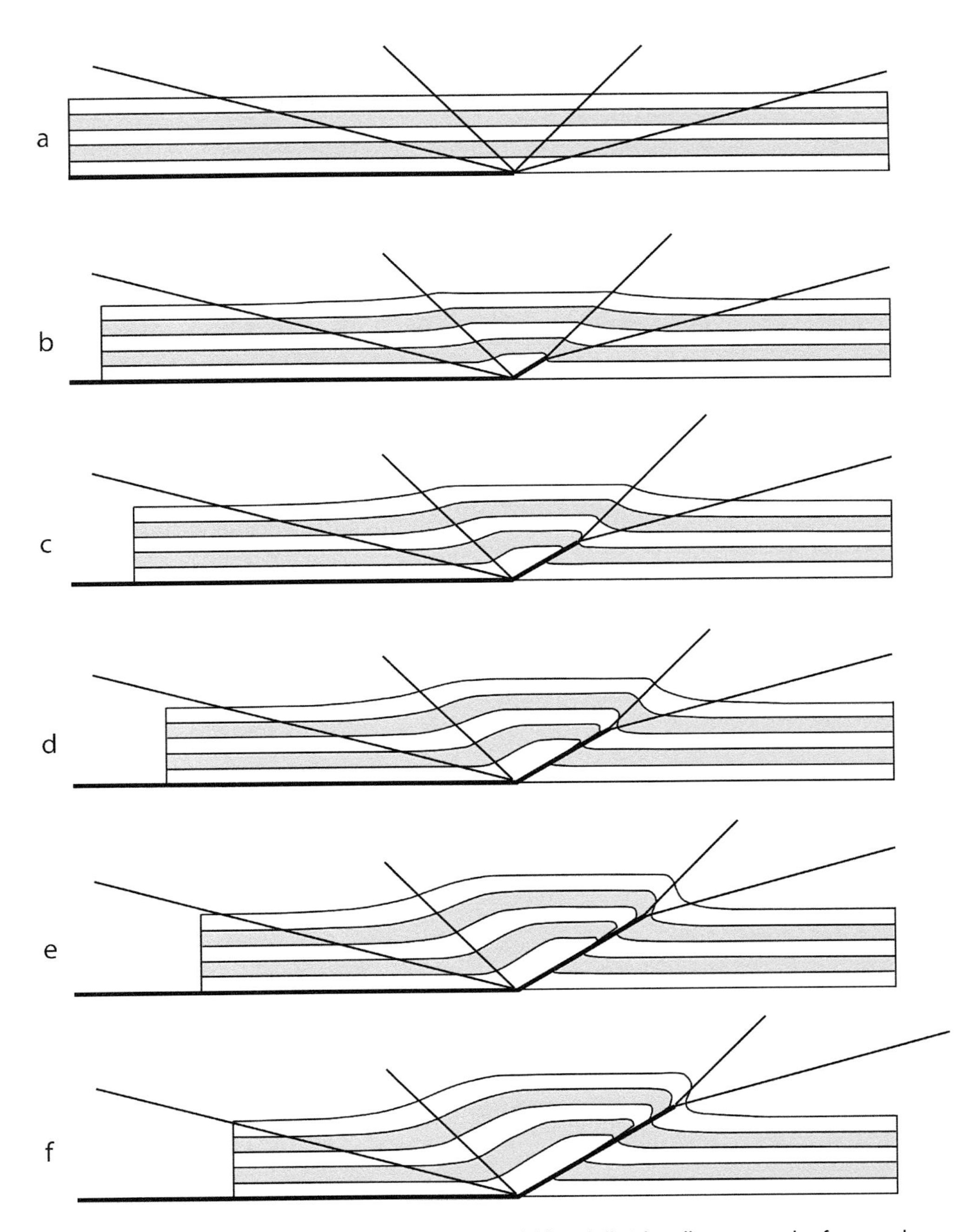

Figure 5.12 Kinematic model of a fault-propagation fold with "trishear" zones at the front and rear of the structure. See explanations in the text. Cross-sections (a) to (f) correspond to the kinematic steps and to some of the cross-sections in our 3D model shown in Figure 5.15.

Source: After Buil (2002)

5.5 3D models of fault-bend folds

We first present three 3D models of fault-related folds (Figs. 5.13 to 5.15) in order to illustrate the kinematic issues developed earlier (Figs. 5.7, 5.9 and 5.12). These 3D blocks are balanced and were created using the profiles presented in Figures 5.7, 5.9 and 5.12. Conservation of the surfaces is thus verified for each of the possible cross-sections between the "initial" non-deformed states and the "final" deformed states constituting the parallel faces of the blocks. The other two faces are convergent, expressing the increased shortening between the "initial" and "final" states. A fourth model is more generic (Fig. 5.16), since its aim is not to illustrate kinematics but rather to show specific geological objects resulting from cutting a structure by the topography.

Figure 5.13 shows a 3D model of a fault-bend fold with four layers, two of which are intersected by the ramp using the kink method. The third dimension makes it possible to understand the lateral evolution of the structure (so it is also a temporal dimension). A valley was cut in the heart of the structure to provide a view of this evolution. The small tectonic window (in blue) at the heart of the structure is a nod to John Rich and to the first fault-bend fold cross-section because it was through such a structure that he came to understand the geometry of the Powell valley (Fig. 5.1).

Figure 5.14 shows a four-layer 3D model of a fault-propagation fold using the kink method. The third dimension makes it possible to understand the lateral evolution of the structure (it is therefore, as in Fig. 5.13, also a temporal dimension). A valley was cut in the heart of the structure to provide a view of this evolution. The ramp dies out in the youngest layer (yellow). The absence of forward slip in the structure explains why the front edge of the model is perpendicular to the two lateral faces corresponding to the initial state and the final state.

Figure 5.15 shows a four-layer 3D model of a fault-propagation fold with trishear zones at the end of the ramp and at the flat-ramp junction. Note the development of a syncline under the ramp. This syncline, like the anticline on the other side of the ramp, is not, as might be expected, a "drag fold," that is, a fold resulting from a twisting of the strata under the action of the fault. On the contrary, as the model illustrates, the development of the fold precedes that of the ramp. The ramp is intersecting a fold that has already formed. A useful analogy is that of the "ductile bead" that develops at the end of a moving fault (Fig. 3.12). The "bead" is expressed here by a fold simply because the ramp propagates by intersecting the strata.

Figure 5.16 is a three-layer model of a thrusting structure, designed as a tool to aid understanding during mapping exercises. A thrust can be observed intersecting a fold. The thrust separates an allochthonous domain from an autochthonous domain. Given the magnitude of the thrust, the allochthon can be described as a thrust nappe. The thrust is in the nascent state on one of the faces but much more evolved on the other (with, therefore, a displacement gradient from one face to the other). Erosion creates a klippe, that is, an allochthonous outlier detached from the main body of the nappe. It also creates some tectonic half-windows, that is to say, areas where the autochthon under the nappe can be observed on the map. Nevertheless, the actual autochthon can be reached (from the half-window) without crossing the nappe. This is why it is a half-window and not a tectonic window. On a map a klippe and a tectonic window are completely surrounded by a thrust. However, the thrust is under the klippe but above the window.

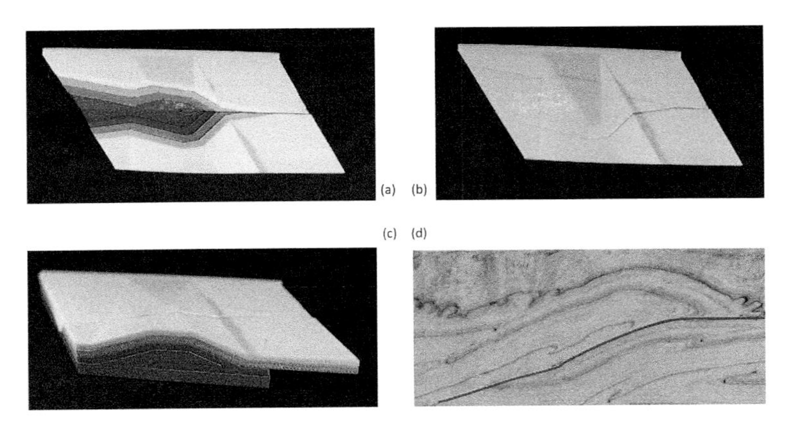

Figure 5.13 A 3D model of a fault-bend fold constructed according to the kinematic model of Figure 5.7: (a) map view, the valley provides a view of the heart of the structure and, in particular, a small tectonic window in blue under the basal stratum in red; (b) map view, but with the valley closed to observe the evolution of the relief; (c) oblique view to see the slip transmitted forward to the hanging wall of the upper flat; (d) the Lisan formation in Israel, a fault-bend fold (the photo has been flipped over so as to be more easily compared to the model). The thrust fault is highlighted in red. This thrust essentially corresponds to an upper flat that folded during the progression of an underlying thrust not visible in the photo. Only the lower part of the segment that tilts toward the left corresponds to a ramp, as indicated by the footwall cutoff observed below. This highlights the fact that it is not the shape and the dip that define a ramp but the position of the cutoffs on the hanging wall and the footwall (see Fig. 5.6).

Source: (d) Photo by © Jean-Claude Ringenbach

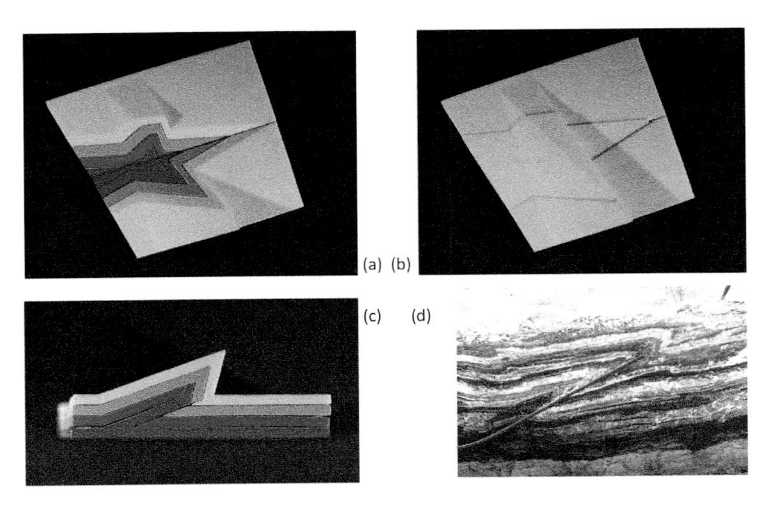

Figure 5.14 3D model of a fault-propagation fold constructed according to the kinematic model of Figure 5.9: (a) map view, the valley provides a view of the heart of the structure; (b) map view, but with the valley closed to observe the evolution of the relief; (c) lateral view to observe the progressive displacement of the strata along the ramp (the slip is at its maximum at the foot of the ramp and minimal at the top); (d) terrain photo (Eocene lignites in the Minervois) showing the termination of a fault in a fold and illustrating the concept of fault-propagation (the photo has been flipped over so as to be more easily compared to model (c)).

Source: (d) Photo © Dominique Frizon de Lamotte

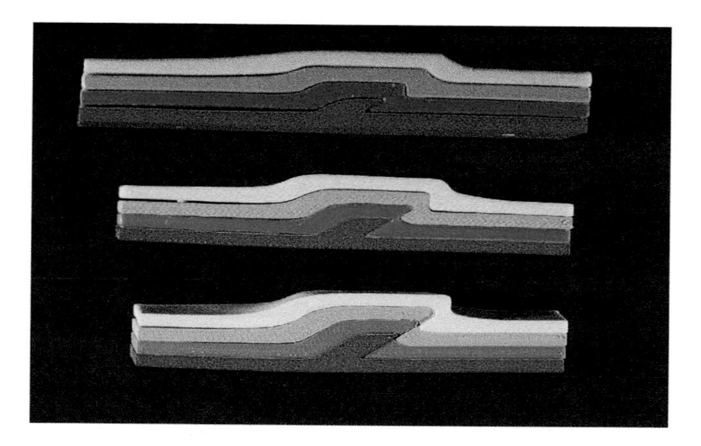

Figure 5.15 Three 3D models of a fault-propagation fold with trishear zones at the front and back of the fold, following the kinematics proposed in Figure 5.13. The three blocks make it possible to view the kinematic evolution of the structure. Note the thickening of the strata at the hanging wall of the ramp as well as in the front limb of the structure (compare the final state (bottom cross-section) with photo (a) of Fig. 5.10).

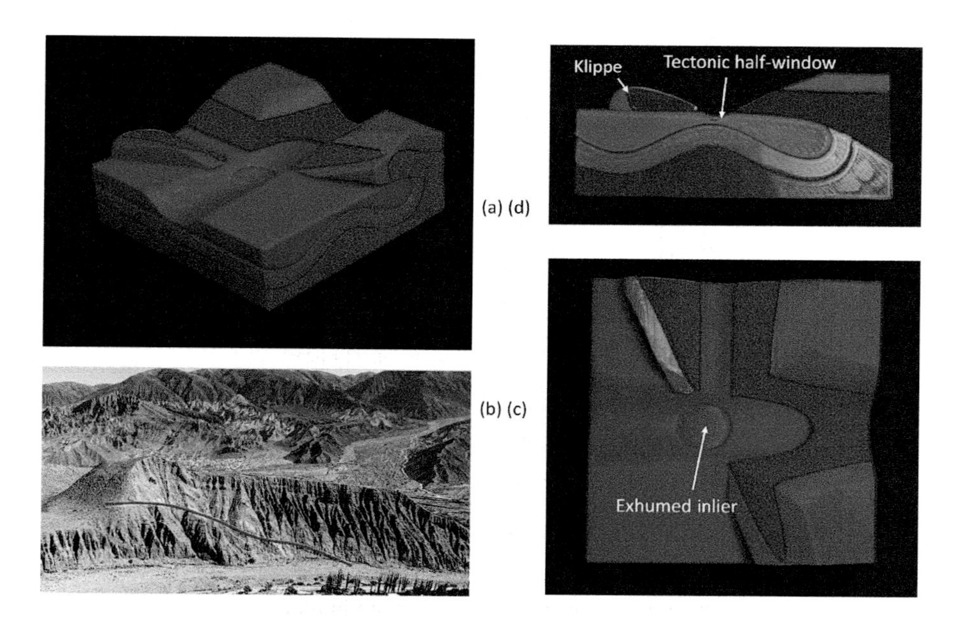

Figure 5.16 A 3D model illustrating the particular features associated with thrust sheets. (a) Oblique view of the model: the thrust fault cuts the model obliquely and places the thrust sheet on its folded autochthon. (b) Terrain photo illustrating the development of a thrust fault putting Pliocene strata above a Quaternary terrace (front of the Andes, Argentina) to compare with the visible side of (a). (c) View from above (on map) of the model; note the nappe outlier (klippe) separated from it by erosion. Between the klippe and the main body of the nappe, the autochthon in blue forms a tectonic half-window. The autochthon in blue reveals the underlying stratum (in purple) by way of a valley. It is an inlier and not a window. (d) Cross-section in the model, allowing comparison with the map view just below. This cross-section corresponds to a lateral photo of the model that has been flipped over for easier comparison, and in particular the sectional expression of the concepts of klippe and tectonic window.

Source: (b) Photo © Jean-Claude Ringenbach

Folding of strata and analog models

The experimental analysis of folds was introduced in the 19th century by James Hall (1815). It was known at the time that sedimentary strata are deposited horizontally. Hall wanted to understand the mechanical origin of the deformations he observed. His experiment consisted in deforming pieces of different kinds of cloth between two vertical wooden boards to obtain folds the same as the ones he observed on the terrain under the effect of a horizontal force. But he still did not understand the origin of this force. Later, Cadell (1887, in Graveleau *et al.* (2012)) wanted to test the hypothesis of thrusts generating major displacements on a kilometer scale by using materials that could not only fold but also break. To test this idea of a "fold-fault," he placed layers of plaster alternating with layers of sand, sand-plaster mixtures or layers of clay to which he applied horizontal compression. His work confirmed that the formation of thrusts can indeed be explained by horizontal forces and that it is possible for them to give rise to displacements over great distances.

Then, in the 20th century, models gradually became analog, with the introduction of the concept of scale. The works of Hubbert (1951) and others have been fundamental to interpret the results in geological terms because they present the physical realism of these experimental models by a precise dimensioning nature – scaling.

Today, there are numerous works using analog modeling, and the associated laboratories have also grown. Imaging techniques as well as knowledge of the mechanics of materials have led to increasingly complex models for which it would be difficult to provide an exhaustive review. Numerous parameters are tested, relating primarily to basal décollement, rear and lateral walls, compression or extension direction, processes and surface kinematics, and structural inheritance. We refer to the remarkable review by Fabien Graveleau et al. (2012).

From a mechanical point of view, models can be built with different materials depending on what is to be modeled: upper crust, full crust or lithosphere. Sand is a good analog material for the fragile behavior of rocks, since it deforms in a localized way with negligible cohesion in relation to the constraints applied in the model. If the model goes down to the lower crust, then a viscous material such as silicone at the base of the model is a good analog material for ductile deformation. This is a simplification but the main aim of the analog simulation is to understand the influence of mechanical and geometrical parameters on the deformation. Thus, many different scale models are possible, but the number of parameters to be tested should be limited.

Analog models that use sand exclusively, thus representing the deformation on the first kilometers of the earth's crust, give rise to a paradox. Deformed models do not really present folds or faults, as can be observed in the field, and yet it is acknowledged that sand is the best material to represent fault-bend folds. In any case, it is the material that is commonly used: its physical characteristics can be controlled and scaled in relation to nature and can, of course, also be varied. Furthermore, it will never be possible to simulate a fault-propagation fold with a granular material such as synthetic or natural sand for a number of reasons. (1) The natural slope of sand – or its internal friction angle – is between 30 and 40°, whereas the measured dips are much larger during the development of a propagation fold. (2) The development of a fault is instantaneous over the whole thickness of the sand of the tested prototype, whereas a propagation fold will propagate over time with the tectonic compression from the basal décollement to the surface. (3) The material transported to the hanging wall of the deformed structure will not be observed in the laboratory, but another fault will be generated at the front of the sand prototype.

It should also be borne in mind that the result obtained by analog simulation is not unique, as was pointed out previously, and depends on many conditions, just like a numerical simulation. A comparison exercise was carried out between the compression analog models of eight laboratories (Fig. A). The prototype to be used was as follows: a 3.5 cm slice of sand composed of, from the base to the surface, 0.5 cm of sand, 0.5 cm of glass beads and 2.5 cm of sand, topped with a triangular prism also made of sand. This pile of sand is subjected to horizontal compression. The eight participating laboratories used their own experiment box and their own sand. The result was observed after 14 cm of shortening, and it varied considerably from one laboratory to another. Results (a), (c), (d), (e) and (f) were qualified as fault-bend folds, and the other two as décollement folds. The following parameters were different: the number and dip of the ramps, their vergence, their location and the thickness of the fault zone. These differences derived primarily from the compression conditions (there are two experimental configurations: a movable rear wall that is pushed or a movable base that is pulled), the dimensions of the box, the view (on the sides or in the center through a medical scanner), the properties of the materials, the slip properties of the base and lateral walls and even the experimenters themselves. As regards the properties of the materials, this can be caused by internal friction, basal and lateral friction (sand/base plate friction and sand/lateral wall friction), density, granulometry (grain size, homogeneity) or cohesion.

We conducted a compression experiment at Cergy-Pontoise along the same lines and as simple as the one described using only sand with the same box and under the same human and material conditions (Fig. B). The only difference was the type of sand used for each experiment, which were two dry sands from Fontainebleau, both non-cohesive but with different granulometry. The one used in the first experiment was homogeneous, while the second was rather heterogeneous. After several centimeters of shortening, a difference in the number of ramps, their vergence and their dip was observed. For the sand with a homogeneous grain size, the faults were irregular and could split into several branches. By contrast, for the sand with a heterogeneous grain size, the faults were more regular and box folds were observed.

These two examples show that experimental conditions have a significant influence on the result. It is therefore important to present statistical results whenever possible. To do this, the experiment must be repeated, under similar experimental conditions if possible, and taking side effects into account. The more these models are controlled, the easier it will be to switch to a numerical simulation. Indeed, the value of such physical modeling lies in obtaining data whose parameters and their variability are known in order to compare them as closely as possible with numerical simulations. It is much more complex to build a numerical model based on natural data because we do not know the laws of rock behavior at geological time scales (see Chapter 2). In structural geology, the rise of analog modeling has allowed us to understand the influence of a number of parameters. The aim is not so much to reproduce a geometry but rather to try to understand the mechanical parameters that influence the formation of a geological structure.

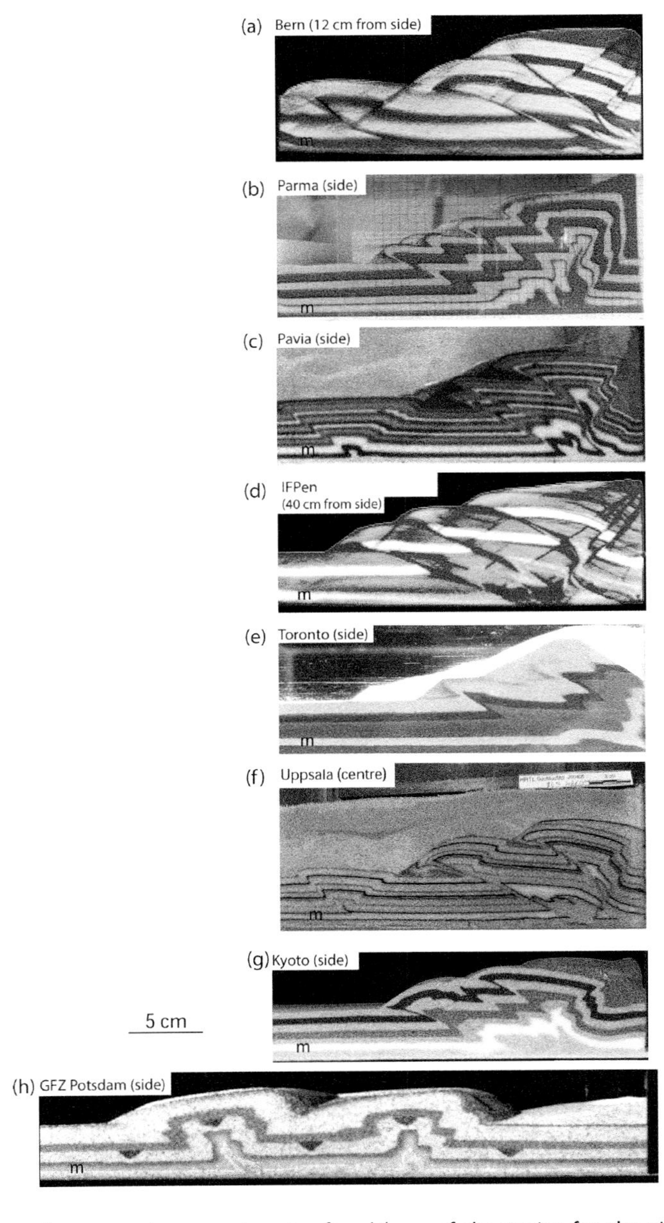

Figure A Views of compression experiments after 14 cm of shortening for the eight participating laboratories whose names are shown in the figures. All the models used dry, non-cohesive sand with a layer of glass beads marked with the letter "m." The Bern and IFPen laboratories used a medical scanner to obtain cuts in the center of the box during the experiment. Models (a) to (f) were constructed with a movable rear wall that was pushed, while for models (g) and (h) the rear wall was fixed to a movable base that was pulled.

Source: From Schreurs et al., 2006

(a)

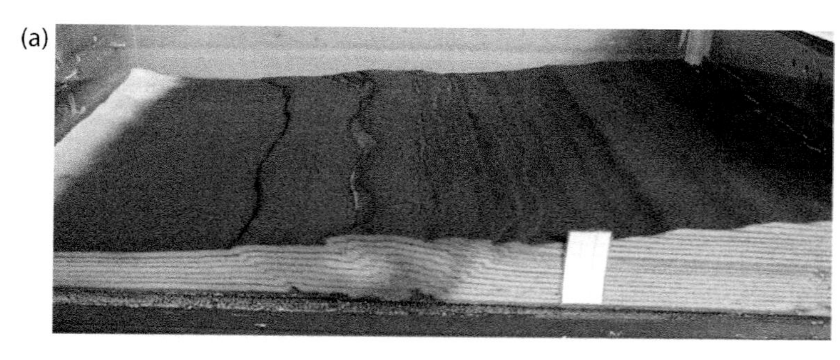

(b)

Figure B Two compression experiments conducted under the same conditions in the Cergy-Pontoise analog laboratory with (a) a sand with homogeneous grain size and (b) a sand with heterogeneous grain size.

Source: Cubas, 2009

5.6 Interaction of individual structures in fold-and-thrust belts

Fault-related folds are rarely isolated structures. They generally combine to form fold-and-thrust belts. In orogenic systems, fold-and-thrust belts constitute what are known as external zones (Fig. 4.19), which differ from the internal zones due to the absence of metamorphism and therefore have a very different tectonic style. There is generally no metamorphism in intracontinental chains (except for any that may have been inherited from the rifting period), so folds and thrusts are therefore the dominant structures.

Understanding a fold-and-thrust belt cannot be reduced into a description of the sum of the individual structures. The conditions of their association in space and time must also be understood. This involves identifying the position of the levels of décollement in the sedimentary pile and especially the chronology of the development of the structures, which is known as the sequence of thrusting. These usually initiate successively from the hinterland to the foreland of the range in a sequence known as forward-breaking. A sequence in the other direction, from the hinterland to the foreland, known as a break-back sequence, could be described but remains anecdotal. Conversely, the development of late thrust faults cutting across existing structures is common, almost to the point of being the rule. Thus, these out-of-sequence thrust faults, that is, thrust faults that do not follow the established sequence, have been observed in many mountain ranges, such as the "frontal pennine thrust" in the Western Alps, which separates the internal zones from the external zones (Fig. 4.19).

The structures present in fold-and-thrust belts include a characteristic structure known as a duplex. This is a stack of thrust slices, known as "horses," bounded by a roof thrust and a floor thrust (Fig. 5.17). Each stack is in turn surrounded by thrusts, and all these thrusts are connected to each other (Fig. 5.17). The sequence of thrusting in a duplex is typically forward-breaking. The first thrust fault to appear is the roof thrust. Then the other thrusts will develop progressively, one underneath the other, by folding the thrusts that appeared at an earlier stage (Fig. 5.17). It is clear in the case of a duplex that any other sequence is impossible. It cannot be imagined that horse labeled 1 on Figure 5.17 forms last because, for that to happen, it would have to go up a very steep slope (Fig. 5.17). This would be unrealistic. Conversely, it can be assumed that thrusting continues, albeit to a lesser degree, during the development of the new basal thrust. The formation of a duplex is accompanied by the creation of a relief, which points to a certain difficulty for the deformation to propagate forwards, and in fact the new structures do not develop forwards but under the previous

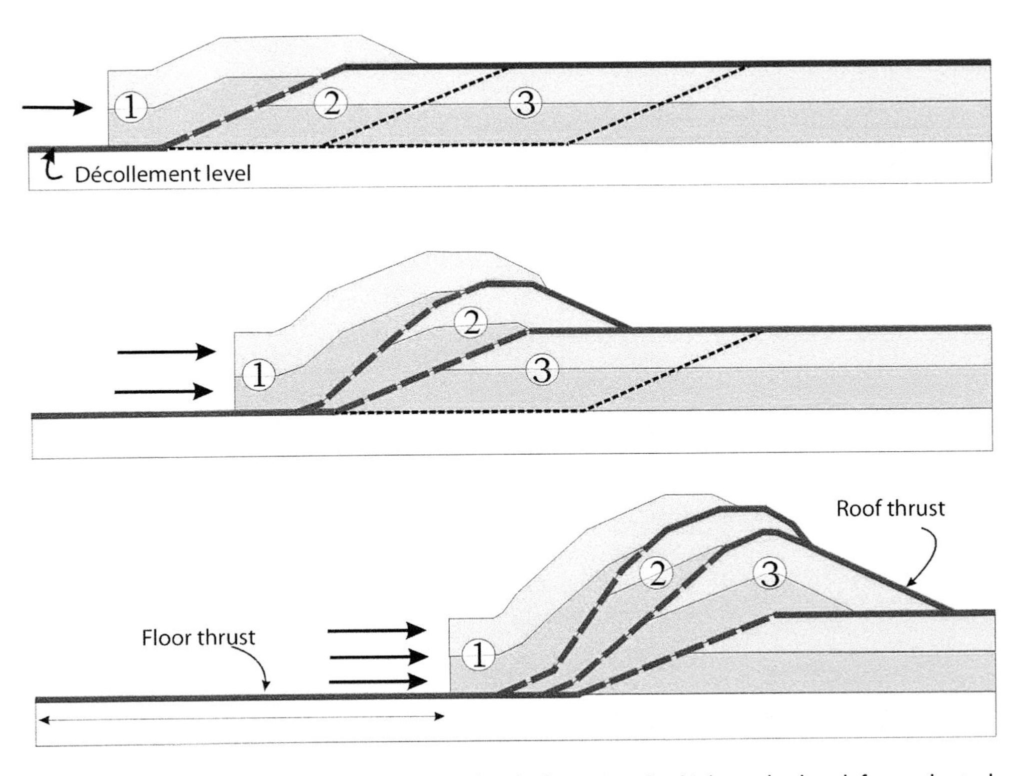

Figure 5.17 Kinematics of the development of a duplex using the kink method and, for each stack, the kinematics of the fault-bend fold (see Fig. 5.7). Each thrust fault develops underneath the previous one. The formation of a duplex illustrates the concept of forward-breaking sequence of thrusting (from the back to the front) and underthrusting (each new stack is formed under the previous one and not in front of it).

Source: Drawing by Christine Souque (2002)

structure. It is said that geological strata go from autochthonous to allochthonous by tectonic under-plating and not by frontal imbrication.

Duplexes of various scales can be seen in mountain ranges, from the outcrop scale to the scale of the portion of the range. Figure 5.18 shows an observable duplex at landscape scale that developed in a restricted part of the sedimentary pile of the front of a fold in the Eastern Pyrenees (France). Crustal-scale duplexes can be seen in many mountain ranges. Thus, in the Western Alps, the external crystalline massifs (Figs. 2.7 and 4.19) constitute a basement duplex and also tectonic windows at the same time. The stacking of thrust sheets by underthrusting leads to the formation of a tectonic culmination, which deforms the overlying nappes favoring their erosion. The highest peaks of the Alps (Mont Blanc) are located in the external crystalline massifs. The concept of the duplex makes it possible to understand that these high summits expose the oldest rocks (Variscan basement), which are also the lowest in the structure.

The Zagros range (Iran) can be used to summarize and illustrate the structures associating folds and thrusts at the scale of a complete orogen. The Zagros is a recent, still active, range belonging to the great family of the Alpine-Himalayan ranges, resulting from the subduction of an oceanic domain (the Neo-Tethys) and then the collision between the Arabian and Eurasian plates. The cross-section shown here (Fig. 5.19) is typical of the eastern part of the range (Lurestan).

Figure 5.18 The "La Cagalière" duplex above the village of Lagrasse (Aude, France). This superb structure is made up of four thrust slices of Eocene limestones ("ilerdian" limestone).

Source: Photo © Dominique Frizon de Lamotte

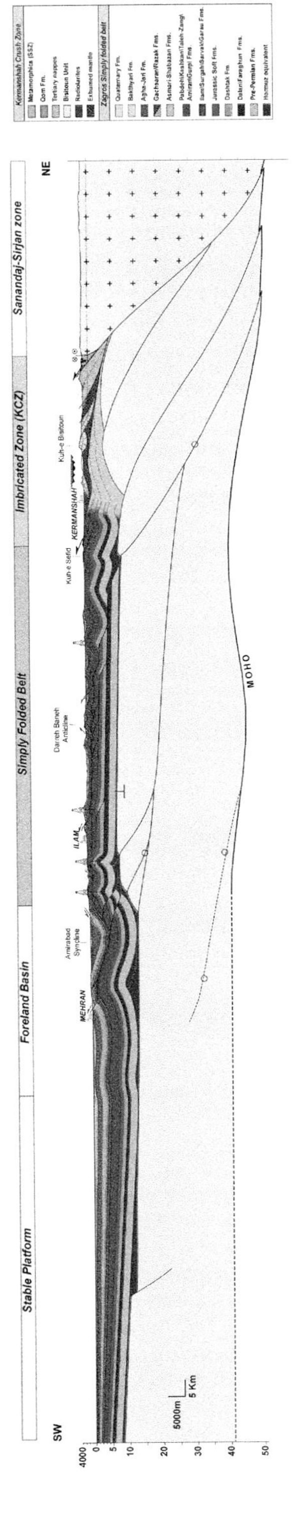

Figure 5.19 Complete cross-section of the Zagros range in the Lurestan region. This cross-section is balanced at the crustal scale.

Source: After Wrobel-Daveau, 2011

On this cross-section, only the northernmost unit (Sanandaj-Sirjan zone) belongs to Central Iran and constitutes the upper plate of the system. Apart from a few outliers of ophiolites highlighting the Neo-Tethys suture, the rest of the system belongs to Arabia, the lower plate of the system. A basement duplex refolding the ophiolite-bearing nappes can be observed in the innermost domains ("imbricated zone"). It should be noted that this crustal duplex does not give rise to large reliefs in the Zagros range. This is most likely because the crust is partly dragged by subduction.

Toward the south, the simply folded belt shows an essentially thin-skinned deformation combining fault-bend folds and décollement folds which developed over the lower décollement (the Hormuz salt formation) on top of the Precambrian basement and with the help various intermediate décollements. Further south, the foreland basin shows a considerable thickness of Cenozoic sediments folded by very large anticlines interpreted as "pop-ups" (see Fig. 3.17). A large basement fault can be observed crossing the entire crust and terminating

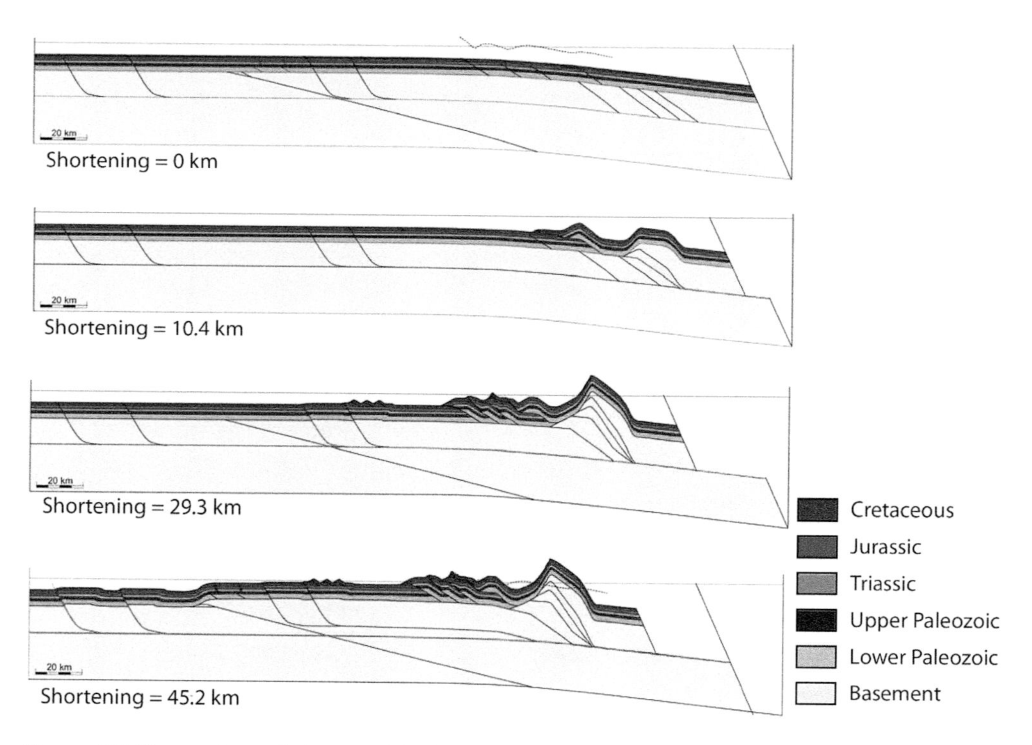

Figure 5.20 Kinematic evolution model of the Zagros range in four stages (after Wrobel-Daveau, 2011). This model was built using GEOSEC software. Stage 1 (16 Ma) corresponds to the initial state and the beginning of the propagation of the deformation in the Arabian platform; Stage 2 (14 Ma) shows the simultaneous development of the propagation of the deformation in the cover and the underthrusting of basement thrust slices (initiation of the basement duplex); Stage 3 (12 Ma) shows the activation of a secondary décollement in the cover and the continuing construction of the basement duplex. Stage 4 (5 Ma) sees the development of the large thrust intersecting the entire crust. By comparison with the geological cross-section (Fig. 5.19), the model of the crustal duplex shows a larger relief. This is due to the impossibility of modeling a continental subduction in a kinematic model when the physical parameters are not known.

without emerging (known as a blind thrust) at the boundary between the foreland basin and the simply-folded belt. This thrust fault was positioned at this location to explain the considerable jump in elevation between the two zones. Its geometry is not entirely satisfactory. What is certain, however, is that the basement is currently involved in the deformation under the Zagros. So it went from a thin-skinned to a thick-skinned style. However, it can be observed that the deformation of the cover is not in line with that of the basement. As in the external Alps, the cover and the basement are decoupled.

A possible deformation sequence is proposed in Figure 5.20. From a non-deformed initial stage, the deformation propagates southwards in a forward-breaking sequence. It should be noted that the range is constructed by combining an underthrusting mechanism in the basement and a frontal imbrication mechanism in the cover. This difference in style explains how the deformation propagates far into the cover, promoting activation of the basal décollement. At the final stage, the basement and cover are coupled again, and the deformation can propagate to the foreland basin and beyond.

Chapter 6

When the salt plays the major role

The effects of salt activity and the concept of salt tectonics

You are the salt of the Earth.

(Matthew 5:13)

6.1 Deposition of salt and geodynamics

Salt (halite, NaCl) and, more generally, the evaporites frequently develop at the bottom of a sedimentary basin and mark the transition between continental deposits (for instance, fluvial sandstones) below and marine deposits (for instance carbonates) above. This is not coincidence: often deposition of salt is associated with the birth of a sedimentary basin, marking a progressive increase of the accommodation space (i.e., the space that is available for sedimentation).

Therefore, salt may be deposited during or just after a rifting, as shown by the margins of the equatorial or southern Atlantic. In this first case, the development of a slope toward the ocean during the post-rift thermal subsidence of the margin (see Chapter 2) can quite rapidly trigger the mobility of salt. Salt can also deposit between two episodes of rifting such as along the margins of the Central Atlantic, in the Atlas system or in the North sea. In this second case, upward salt movement is stimulated during the post-salt rifting episode and can continue during a long period, giving birth to spectacular structures like salt glaciers (more on this later). Salt may also be deposited at the initiation of a flexural basin in front of an active orogenic system. It is the case of the "Gasharan" Miocene salt in the Zagros mountains (see Figs. 5.19 and 5.20). Interestingly, the Zagros displays two major salt levels: the rift-related Cambrian "Hormuz" salt and the Miocene "Gasharan" salt marking the onset of the bending (flexuration) of the Arabian lithosphere.

From the point of view of applied geology, the exceptional physical properties of salt and, in particular, its very low permeability, explain why it is an excellent seal for natural oil and gas reservoirs. For the same reasons, the evaporites also represent good sites for the storage of gas/oil or even for nuclear waste. Interestingly, it is easy to create a cavity for storage in salt by solution mining, that is, extracting salt by dissolving this mineral with water.

6.2 Definition of the different types of salt-related structures

Diapir is a generic term designating a ductile mass penetrating the overlying brittle rocks (Fig. 6.1). If the overlying strata are thinned but not pierced, the term salt dome should be preferably used. Salt is not the unique material giving birth to diapirs: evaporites in general,

Figure 6.1 Example of diapir, to the left, penetrating sedimentary rocks on the right, Sivas basin (Turkey).
Source: Photo © Jean-Claude Ringenbach

Figure 6.2 (a) Salt plug and (b) salt wall; Key: gray: basement; pink: mobile layer (salt); blue: pre-kinematic sedimentary layer; orange and yellow: syn-kinematic layers.

clays (the concept of mud-diapirs) or even, in another context, igneous intrusions (granite) can play the same role. However, salt-related diapirs are certainly the more spectacular; in general, halite is not the unique component of salt: it also comprises anhydrite, gypsum and clay minerals.

The salt diapirs exhibit different shapes and can be classified according to two end-members: plugs and walls. A salt plug (or salt stock) presents a subcircular shape in map view (Fig. 6.2). By contrast, a salt wall (or salt ridge) presents an elongated shape. A salt-cored anticline, like in the Jura, is not strictly speaking a salt wall because the salt does not pierce

Figure 6.3 An example of an active salt glacier flooding in a valley of the Fars province (Eastern Zagros, Iran).

Source: Photo © Dominique Frizon de Lamotte

the overlying strata (see Fig. 4.7). However, in the fold-thrust belts (Zagros is the best example), salt-cored anticlines frequently correspond to ancient salt-walls reactivated during the tectonic inversion. Frequently, salt plugs merge at depth with salt walls or develop at the intersection of salt walls.

A salt glacier is a flow of salt that forms when a rising diapir reaches the surface of the earth and floods along this surface. The name salt glacier highlights the similarities with ice glaciers. The Fars region in the southeastern part of the Zagros mountains (Iran) expose spectacular active salt glaciers (Fig. 6.3). Their exceptional preservation is due to the arid climate there. In the sedimentary basins, ancient salt glaciers, sometimes called salt tongues, can be preserved (sealed) by overriding sediments covering them rapidly.

In this context, the coalescence of a number of glaciers gives birth to a salt canopy. Salt canopies buried at depth have been recognized in several sedimentary basins and continental margins such as the Gulf of Mexico or the Central Atlantic margins (Fig. 6.4). In these two examples, the salt is Upper Triassic in age. Onshore, the so-called Tell-Rif nappes in North Africa expose ancient submarine salt canopies (Fig. 6.5). Some of them acted as décollement levels during the Cenozoic tectonic inversion of the former North African margin, now imbricated in the Tell-Rif orogenic system (Fig. 6.5).

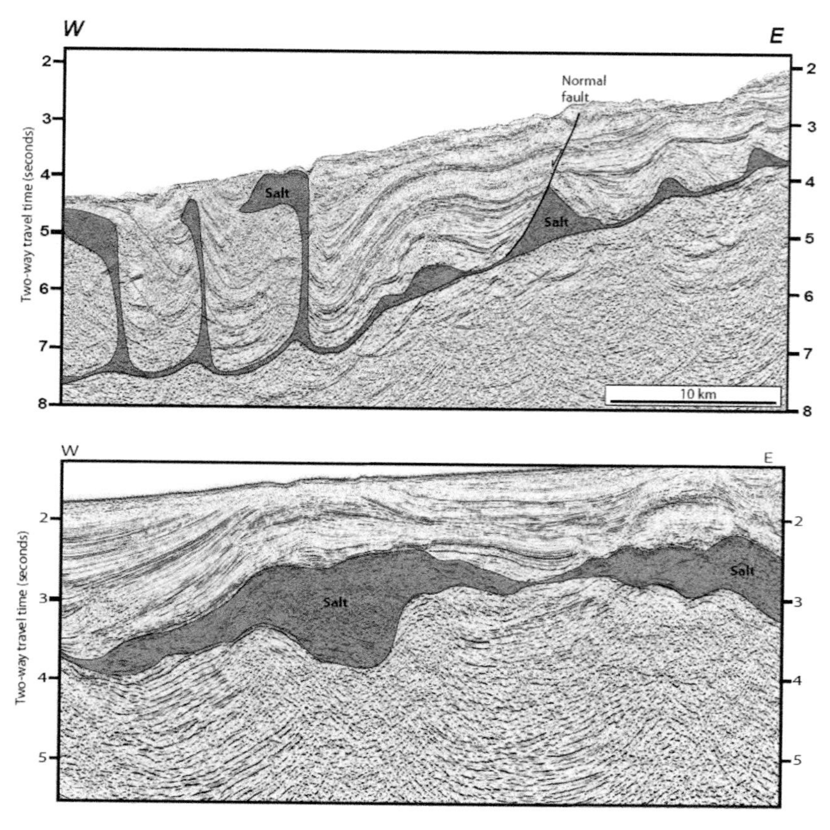

Figure 6.4 Examples of salt-related structures along the Atlantic rifted margin of Morocco. Top: salt tongues and salt walls; bottom: salt canopy.

Source: Modified from Tari and Jabour (2013)

Figure 6.5 Former salt canopy exposed in the Aknoul nappe (external Rif, Morocco). The yellow line underlines the limit of the glacier. The salt is Triassic in age; the host rock is made up of Upper Cretaceous marls.

Source: Photo © Dominique Frizon de Lamotte

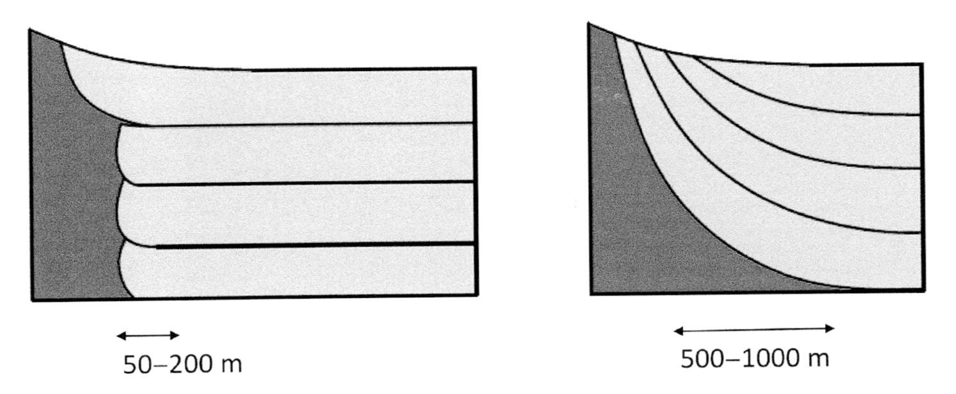

Figure 6.6 Conceptual models illustrating the geometry of hook (left) and wedge (right) halokinetic sequences. Note that the scales of the structures are different.

Source: Modified from Giles and Rowan (2012)

Figure 6.7 Left: field example of a hook halokinetic sequence in the Sivas basin (Turkey). Right: field example of a wedge halokinetic sequence in the Qhesm island (Iran). The yellow dotted lines underline the bedding; the violet dotted lines underline the borders of the diapirs.

Source: Left: photo © Jean-Claude Ringenbach; right: photo © Dominique Frizon de Lamotte

The timing of salt activity can be recorded by specific syn-sedimentary structures defining halokinetic sequences. The authors recognize two end-member types of sequences: the hooks and the wedge halokinetic sequences, respectively (Fig. 6.6). The hook sequences are characterized by tabular boundaries of each sequence, with a narrow zone of thinning close to the diapir (Fig. 6.7). The active diapir remains very close to or at the sea floor at each step of the evolution. The occurrence of debris flows of reworked material signs the emergence of the diapir. The wedge sequences show convergent boundaries and a broad zone of thinning toward the diapir (Fig. 6.7); they do not necessitate a breaching of the diapirs at the surface. Different types of sequence can develop through time about a given diapir.

6.3 Causes of salt mobility

For decades, salt mobility was considered as the direct result of a change in the density gradient between the salt and the overburden. Such a shift occurs at about 1000 m because, in contrast to other rocks, the salt density slightly decreases with burial, from 2200 to 2100 kg/m^3

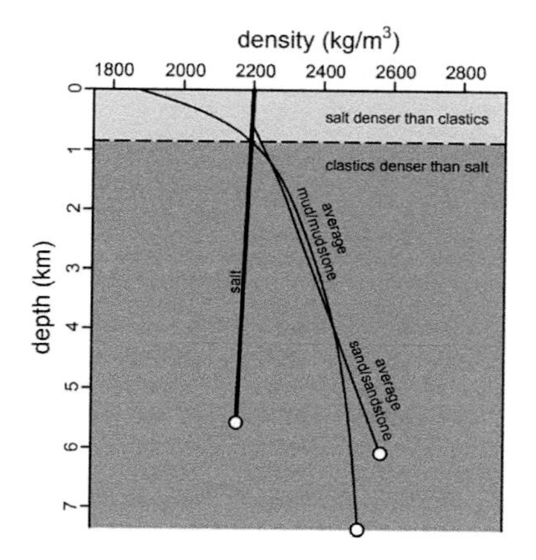

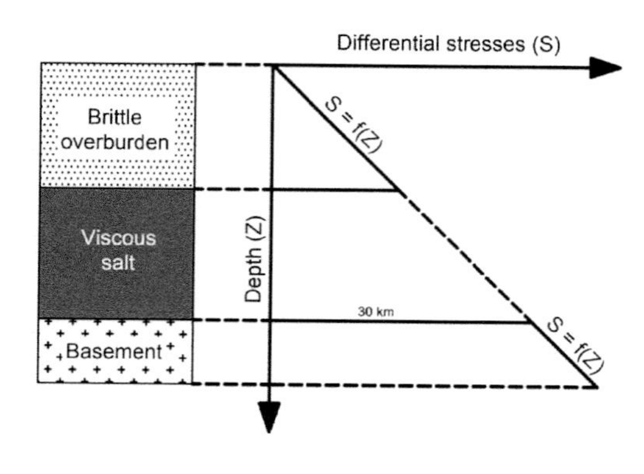

Figure 6.8 Physical properties of salt. Left: salt is buoyant because it does not record compaction with depth; right: salt is weak and present a viscous behavior compared to the brittle (frictional) overburden and basement.

(Fig. 6.8). Another proposed mechanism is the differential loading resulting from an irregular sedimentation above a salt level. In this case, the idea is that the very mobile salt immediately accommodates a difference of load. This process does not require a lighter density of salt compared to overlying sedimentary rocks. The mechanism of mobility depends on the low viscosity of salt and not its low density (Fig. 6.8).

These two mechanisms, linked to density and viscosity of salt respectively, decidedly contribute to its mobility. However, since the 1990s, it is recognized that an external factor is required to trigger salt movement. This additional factor can be related either to tectonic events (i.e., extensional deformation) or to gravity instabilities, or to a combination of both.

6.3.1 Salt mobility and extensional tectonics

The analysis of salt basins together with scaled analog models show that regional extension may trigger the growth of plugs and walls at the expense of a pre-kinematic salt layer. Two main configurations are recognized depending on the symmetry versus the asymmetry of the system of faults. It is worth noting that the structures can develop in a thin-skinned context (as presented in Figs. 6.9 and 6.10) but in a thick-skinned context as well (see the model in Fig. 6.18).

Symmetric systems show no rotation of the pre-kinematic strata but the development of a graben with a diapir (with a triangle shape) filling the space created in its core. Synchronously, syn-rift sediments fill up the space between the extensional faults and the diapir. After the end of the extensional deformation, the diapir can continue to growth "passively" with development of hooks along its flanks (Fig. 6.9). The complete depletion of the salt layer signs the death of the diapir (not shown on the figure).

The asymmetric system allows the rotation of pre-kinematic beds up to a steep attitude on one side of the salt-wall and a roll-over anticline on the other side. Syn-kinematic growth strata exhibit very different geometries on both sides (Fig. 6.10). A nice example of such an asymmetric

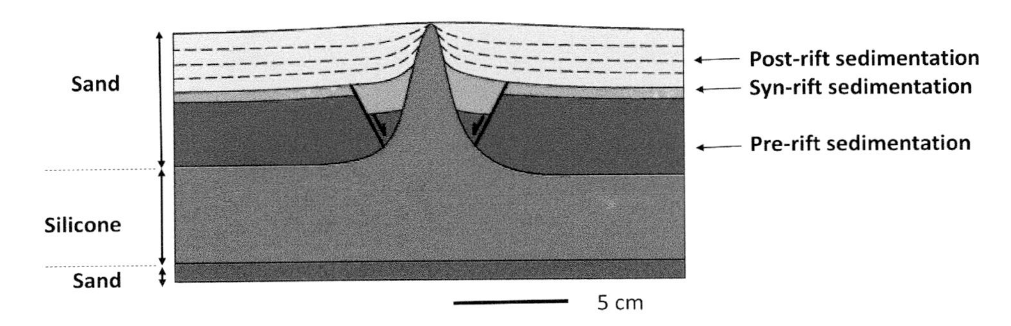

Figure 6.9 Drawing of an analog experiment illustrating salt mobility during a pure shear thin-skinned extensional deformation. The pink layer (representing salt) and the blue layer (representing carbonate) were deposited before deformation. Note the development of hooks along the diapir during the post-rift sedimentation.

Source: Modified from Brun and Fort (2008)

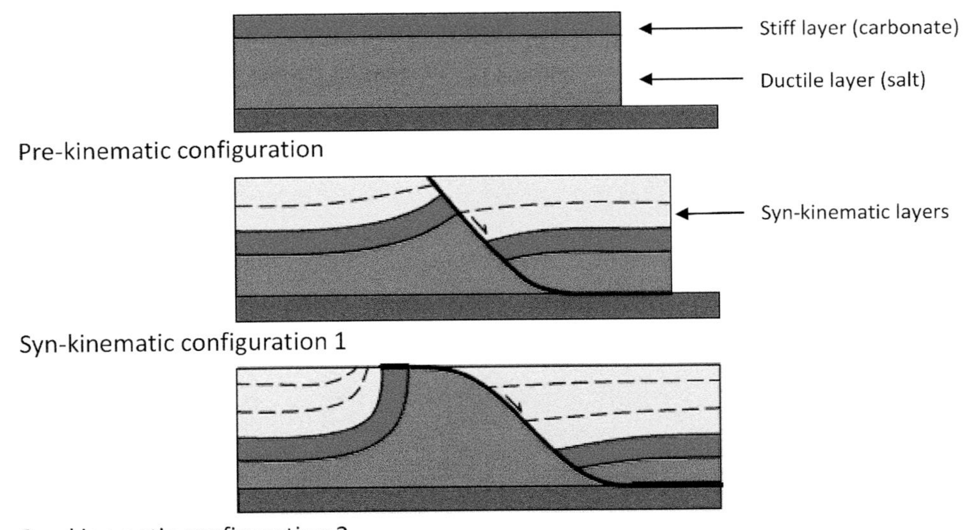

Figure 6.10 Conceptual model illustrating salt mobility resulting from the development of an asymmetric extensional system.

diapir is exposed in the external Rif (Morocco). However, only the side corresponding to the vertical pre-kinematic beds and associated growth strata are well preserved (Fig. 6.11).

6.3.2 *Salt mobility and gravity gliding/spreading*

Post-rift salt is frequent along the so-called rifted (or passive) margins. In this particular setting, salt tectonics is a gravity-driven process in which two factors contribute: gliding and spreading. The gliding is due to the progressive tilt of the margin during the post-rift

Figure 6.11 Photo (a) and interpretation (b) of the Sof Aouzzaï (external Rif, Morocco). This site shows Middle Jurassic growth strata developed over a moving diapir of Triassic salt. The Lower Jurassic carbonate is pre-kinematic. The Upper Jurassic turbidites are post-kinematic. Compare with the left part of "configuration 2" on Figure 6.9.

Source: Modified from Gimeno-Vives *et al.* (2019)

period as a consequence of the cooling of the oceanic crust and thermal reequilibation. It is worth noting that a slope of 1° in a basin hundreds of kilometers wide is enough to trigger gliding. The spreading is driven by the differential sedimentary loading linked to the progradation of sediments oceanward. Even if these two mechanisms are theoritically disctinct and are described by different models (Fig. 6.12), in reality, they play together along the rifted margins, allowing salt movement downward, that is, toward the deep basin.

The organization of the salt-related structures combines updip extension and downdip compression. Both extension and compression develop in a thin-skinned style and are exactly balanced explaining why a "pure" thin-skinned deformation is possible in this specific case (Fig. 6.12). The downdip system looks like a fold-and-thrust belt and can associate detachement and ramp-related folds as described in Chapters 4 and 5. The development of a true "salt nappe" can be inferred in the case of "pure spreading" (Fig. 6.11b). The main difference with a classical fold-thrust belt (like the Jura, Fig. 5.7) is the wide occurrence of growth-strata developed during the gliding over the salt.

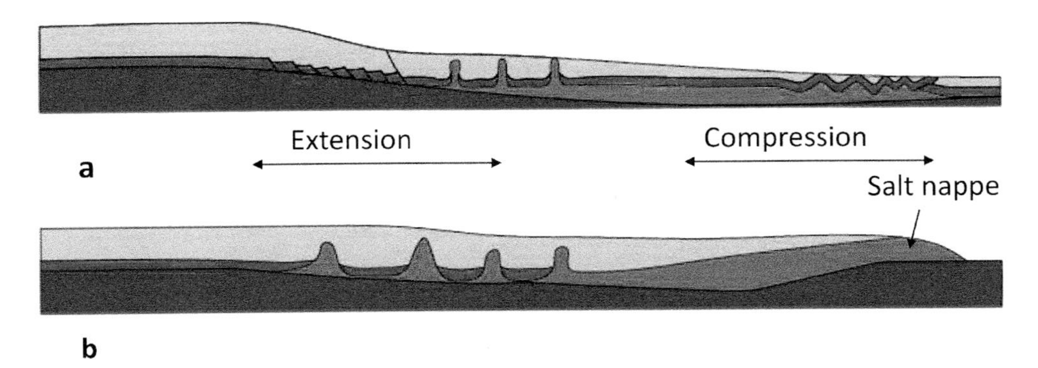

Figure 6.12 Conceptual models explaining the effects of (a) "pure gliding" and (b) "pure spreading.". Note, in both cases, the combination of frontal contraction to the right with updip extension to the left. The blue sediments are pre-kinematic, while the yellow sediments are syn-kinematic relative to salt movement.

Source: Modified from Brun and Fort (2011)

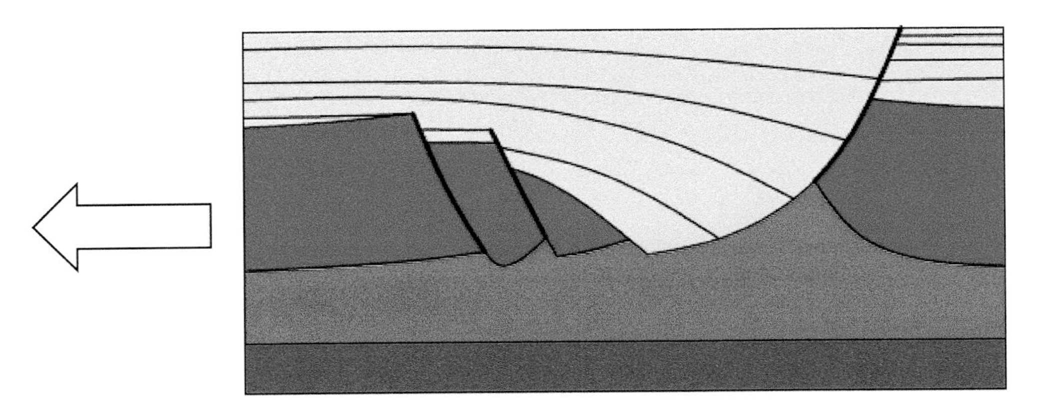

Figure 6.13 Conceptual model explaining the development of a "roll-over" structure filled by growth strata (in yellow) in between two "rafts" (in blue).

The updip system exposes large "rafts" bounded by extensional faults on one side and by a bended structural surface (roll-over anticline) on the other side (Fig. 6.13). Syn-kinematic sediments show growth-strata with a typical wedge geometry filling the space between two rafts. The faults are usually dipping toward the ocean but the counter geometry also exists. When both are combined, they can give birth to "turtleback" structures (Fig. 6.14). A large amount of extension is required for the development of such structures.

In the compressional domain at the foot margin, the folds are frequently localized by former salt walls inherited from the early evolution stages (Fig. 6.15). As indicated previously, such a configuration is also frequent in fold-thrust belts like the Zagros (Fig. 6.16), the Jura (Fig. 4.7) being a counterexample.

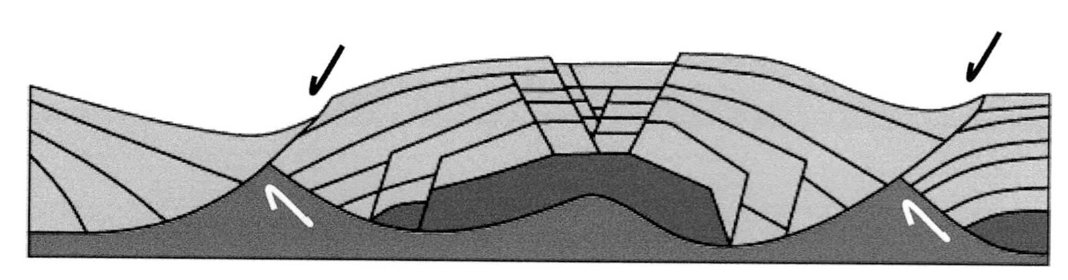

Figure 6.14 "Turtleback" structure. The pre-kinematic "raft" (in blue in the center) is embedded within syn-kinematic growth strata (in yellow) on both sides.

Source: Modified from Brun and Fort (2008)

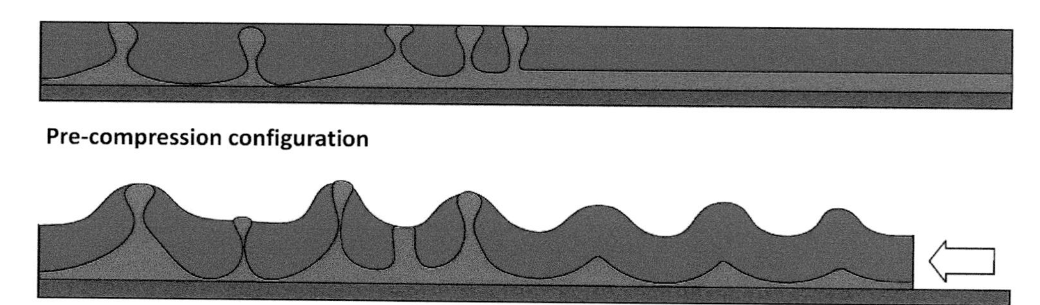

Post-compression configuration

Figure 6.15 Conceptual model explaining how former salt walls can localize folds in a subsequent contractional setting.

Source: Modified from Brun and Fort (2011)

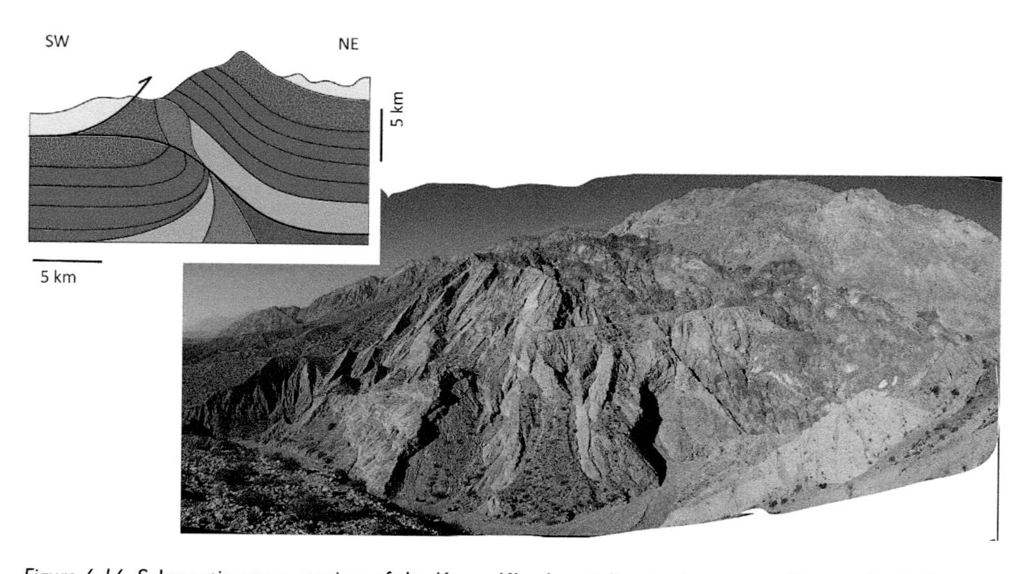

Figure 6.16 Schematic cross-section of the Kue-e-Khush anticline in the eastern Zagros (Iran) illustrating the breaching of a salt plug in the core of the anticline. Growth strata in the Paleozoic (brown color) and Mesozoic (blue color) show that the diapir was already active at that time.

Source: Modified from Sherkati *et al.* (2005) and photo © Dominique Frizon de Lamotte

6.4 Printable 3D models of salt-related structures

Three 3D printed models are proposed to illustrate specificities of salt tectonics. The first model (Fig. 6.17) illustrates generic configurations: a salt wall, a salt dome and a salt diapir. Growth strata fringe all these structures. The salt dome does not breach the above-lying strata. Around the diapir, "flap" structures, formed by overturned strata, are shown. Such structures result from the rotation of the vertical flanks of the diapir as salt breaks out and flows laterally overhead.

The second model (Fig. 6.18) illustrates the development of a symmetric diapir above an extensional basement fault. The salt is pre-kinematic (pre-rift), the above-lying strata are syn-kinematic and syn-rift (violet color) or syn-kinematic and post-rift (yellow color).

The last model (Fig. 6.19) illustrates the development of a rollover anticline above a salt-top extensional fault in a thin-skinned tectonic context. An asymmetric diapir is represented in the footwall of the fault mimicking the configuration on the second stage of Figure 6.10. Growth strata, represented in the anticline-top basins by a fan within the yellow color, are placed to suggest their syn-kinematic nature.

Figure 6.17 Salt diapir with overturned flaps (left), salt dome (center) and salt wall (right). The salt is represented in gray; all the above-lying strata are syn-kinematic layers.

Figure 6.18 Diapir (in gray) developed above an extensional basement (in red) fault. Note the representation of hooks fringing the diapir. On the lateral view, the thickening of the violet color toward the diapir is suggesting that the salt was only mobile vertically during the deposition layer. By contrast, the absence of change in the yellow layer thickness (except along the diapir) is suggesting that salt is also mobile laterally.

Figure 6.19 Asymmetric diapir in the footwall of a listric extensional fault. The salt, in gray, and the above-lying violet layer are drawn as pre-kinematic. The yellow mini-basin, with marked growth strata, is believed to be syn-kinematic.

Appendix

Practical tips for designing and printing 3D objects

As we have indicated in the Foreword to this book, 3D vision is essential for learning geology. This shared observation has led many teachers to tinker with objects made of clay or by folding and gluing cardboard sheets for educational use. 3D printing obviously facilitates this process, and the proliferation of "fab labs" or "fablabs" (contraction of the English "fabrication laboratory") is an obvious accelerator. Some fablabs are located within university campuses, others are designed as innovation centers supported by local authorities and others are mixed and combine the two previous categories. This is the case of the "Labboite" fablab (www.labboite.fr/) in the heart of the town of Cergy-Pontoise, where we made our first prototypes. In general, these "fablabs" are ideal places for developing projects with students.

To make a prototype, it must first be designed. Geologists' field notebooks are full of 3D diagrams (Fig. A). But these drafts, which are essential, are not enough to go directly to the printing of a model. A computer-aided drawing step is required.

It is a matter of producing a drawing that can then be saved in a format that can be printed. All these operations are carried out using specific software. For drawing we used Autodesk® Fusion 360 software, through which we can draw and display the 3D design in a compact or exploded version, that is, where each part of the model is shown individually. At this stage we obtain files in an ".stl" format. The next step is to prepare the 3D images in order to make them ready for printing. This step is performed using the free Ultimaker® Cura software, which can be used to preview and size the object. All these operations are carried out on a computer (Fig. B). The output file is saved on an SD card in a ".gcode" format that can be read directly by the printer.

We are then ready to print (Fig. C). There are an increasing number of manufacturers and distributors on the market with very different prices. We recommend that you contact a supplier who can provide maintenance. The filament used for printing is PLA (polylactide), a plant-based biodegradable bio-plastic made from corn starch or other vegetable starch.

On the printer tray, the printed objects are surrounded by a narrow thin border ("brim") that stabilizes the model during printing. This edge must then be removed with a deburrer (Fig. D). In addition to the printer, a small workshop with basic tools – vice, deburrer, cutter – therefore needs to be provided. Each colored part of a model is printed independently. To obtain the finished model, the parts must be attached to each other. This can be done permanently by gluing the parts with a suitable glue or temporarily by using the holes provided in the parts to fit them using fluted wooden dowels of appropriate size (usually 6 mm). This second solution offers the possibility of disassembling the models and observing the 3D geometry of each part independently. Maintaining this possibility is very useful when

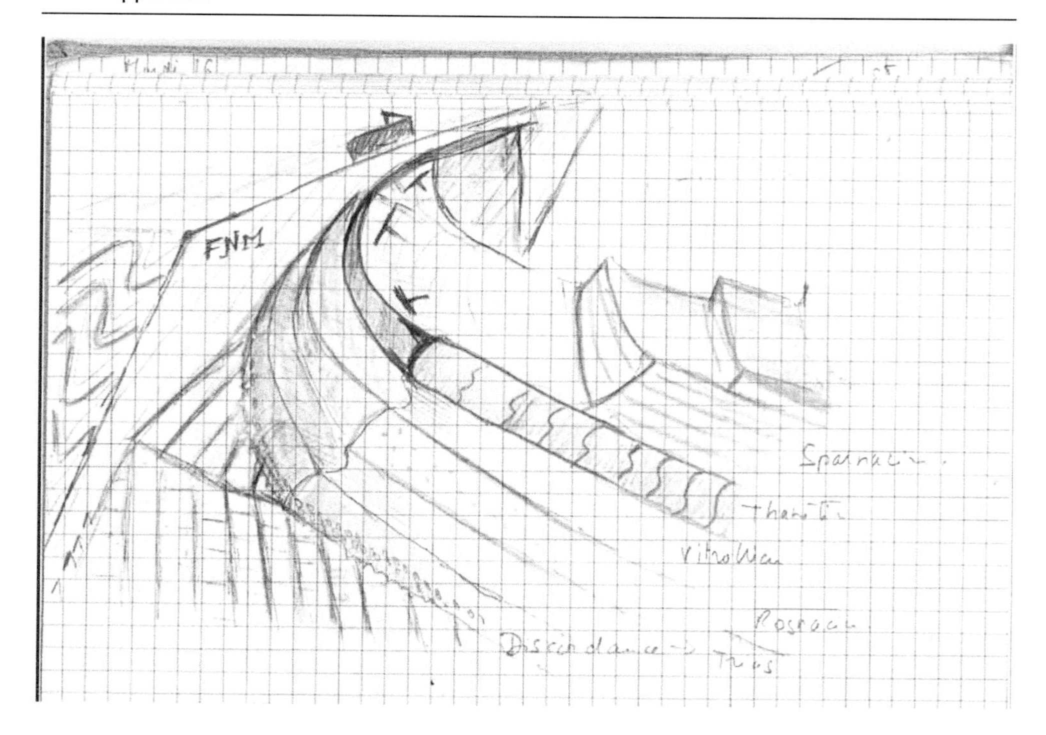

Figure A Example of a 3D diagram taken from a geologist's field notebook (Dominique Frizon de Lamotte in this case). The diagram represents a structure in the Corbières region (Eastern Pyrenees) and is intended to illustrate the effect of the "Nord-Mouthoumet" fault (FNM) on the geological strata against its footwall. Note also the unconformity ("discordance" in French) between the Upper Cretaceous strata ("Rognacian," in green) and the vertical Triassic strata.

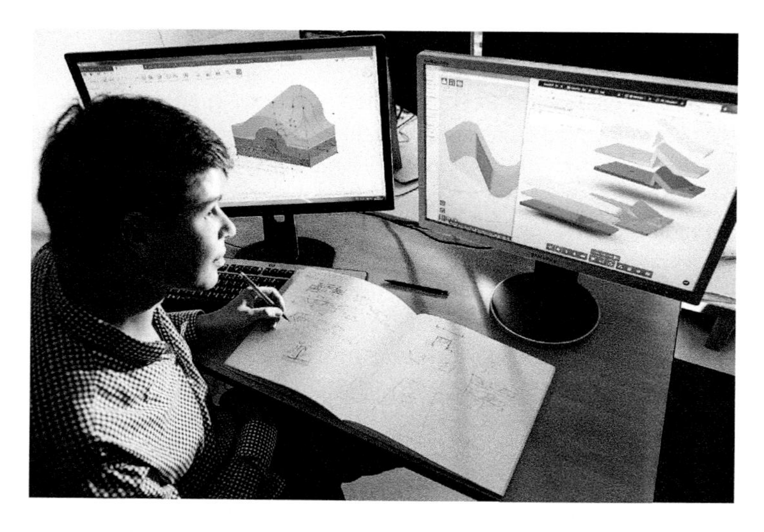

Figure B An operator in front of his screen showing all the stages in a project: in the notebook, the initial drawing by hand; on the left-hand screen, the drawing made using Autodesk® Fusion 360 (drawing of a detachment fold); on the right-hand screen, to the right is the exploded model of a ramp propagation fold displayed using Autodesk® Fusion 360; to the left, the pre-printing display of a similar fold using Ultimaker Cura.

Source: Photo © Alexis Chézière

Figure C Printing in progress. The printer ("scalar" model) is offered as a kit by the company 3D Modular Systems. In the far right foreground are filament bobbins and already printed "kink" pleat patterns.

Source: Photo © Alexis Chézière

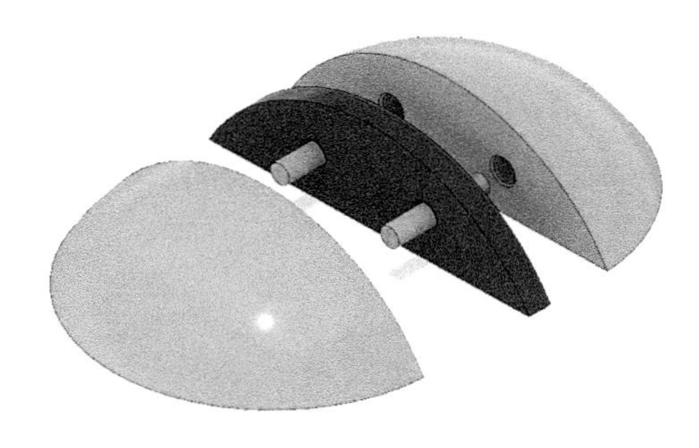

Figure D Left: work in the workshop to finish off a printed part using a deburring machine. Right: assembly system for the model of a small hill with a vertical layer (see Fig. 1.11).

the third dimension provides an opportunity to address the kinematic question, as with the models shown in Figures 4.16, 4.17, 4.18, 5.13, 5.14 and 5.15. In this situation it is in fact essential to have a view of the lateral evolution of each element of the model.

Assembly procedures are shown in Figure D. The model in the figure and some others can be downloaded free of charge from Dunod's website (www.dunod.com) and from the website of the *Société Géologique de France* (www.geosoc.fr), where a catalogue of digital files is also available.

Toward 3D models representing real examples

In this book only models designed to illustrate either a concept (for example an unconformity) or idealized objects (a ramp propagation fold for example) are presented. A complementary approach is to look at real structures. We have tested this possibility on two examples near our university, with encouraging results.

The first model represents the Pays de Bray inlier, close to Beauvais in France (Fig. E). This model can be used to discuss several tectonic questions – folding, faulting, periclinal termination, axial plunge, etc. – as well as geomorphological ones – the formation of an anticlinal valley, a cluse and an exposed core. This is a useful teaching tool before, during and after a geological field trip or during a mapping session.

The second model (Fig. F) represents the loops in the river Oise where the city of Cergy-Pontoise stands. This model can be used to discuss the geomorphology of flat regions and

Figure E Model of the Pays de Bray printed in 3D and placed on the 1/50,000 geological map of Beauvais published by BRGM (*Bureau de Recherches Géologiques et Minières*). The model has the same horizontal scale as the map. The vertical scale is multiplied by 4. Stratigraphy has been simplified and is limited to Upper Jurassic or more recent rocks. Recent formations have not been shown in this model for study of the overall structure and geomorphology, but these elements could be added.

Source: Photo © Pascale Leturmy

Figure F Model of the Cergy-Pontoise site printed in 3D and placed on the 1/50,000 geological map of Pontoise published by BRGM. The model has the same horizontal scale as the map. The stratigraphy has been simplified and is restricted to the Upper Cretaceous (green) or Cenozoic (orange, yellow and purple) rocks. The purple high ground of "L'Hautil" can easily be seen on the pink model on the map in the foreground between the Oise on the right and Seine on the left. The vertical scale is multiplied by 4. Recent formations highlighting the loop in the Oise and the Oise-Seine confluence have been printed in white and placed over the other pieces of the model. Old alluvial deposits and plateau silts have not been shown at this stage. They could be.

Source: Photo © Pascale Leturmy

the dynamics of incised valleys. Here, too, it provides a useful teaching aid for field trips or classroom map analysis sessions.

Other applications in the field of earth sciences

In this book, because of our academic specializations, we have focused on structures of tectonic origin. But it is not the only field in earth sciences where 3D printing is useful for teaching.

Among the very many conceivable topics, the visualization of successive deposition sequences during sediment filling at a continental margin is a massive 3D problem. In fact eustasy, that is the variation in absolute sea level, together with sedimentary inputs and subsidence, will control the space available for sedimentation, which is called the accommodation space. Thus, sequences separated by discontinuities (air or possibly marine erosion surfaces) corresponding to gaps in sedimentation (hiatuses) follow one another in time and space.

Figure G A 3D-printed five-layer sequential stratigraphy model (see Fig. H for element details). The discontinuities correspond to the changes in color.

Source: Photo © Alexis Chézière

These sequences, in which geometry varies transversely and longitudinally, can be used to establish a sequential stratigraphy. Our model (Fig. G) illustrates a typical stack with, from bottom to top:

- a high stand system tract, which on the model corresponds to the red basement (Fig. G and Ha);
- an underwater cone in purple on the model (Fig. Ha) following a major regression;
- a low stand system tract in blue on the model (Fig. Hb) reflecting the start of a transgression;
- a transgressive system tract in green (Fig. Hc and Hd) marking the climax of the transgression;
- another high stand system tract completing the cycle (Fig. Hc).

Sediment filling takes place in two ways depending on the geometry of the accommodation space. It may take place by aggradation (vertical stacking of horizontal layers) or by progradation (lateral stacking of inclined layers reflecting the lack of space for vertical stacking). On the model (Fig. H), progradation is marked by oblique black lines (visible on the red and orange sequences), while aggradation is marked by horizontal black lines (visible on the purple sequences and the upper part of the red part). These geometries depend directly on growth or decrease in the accommodation space, which itself will ultimately depend on eustasy. Thus, aggradation indicates a large space available on the platform, while progradation indicates a small space available on the platform.

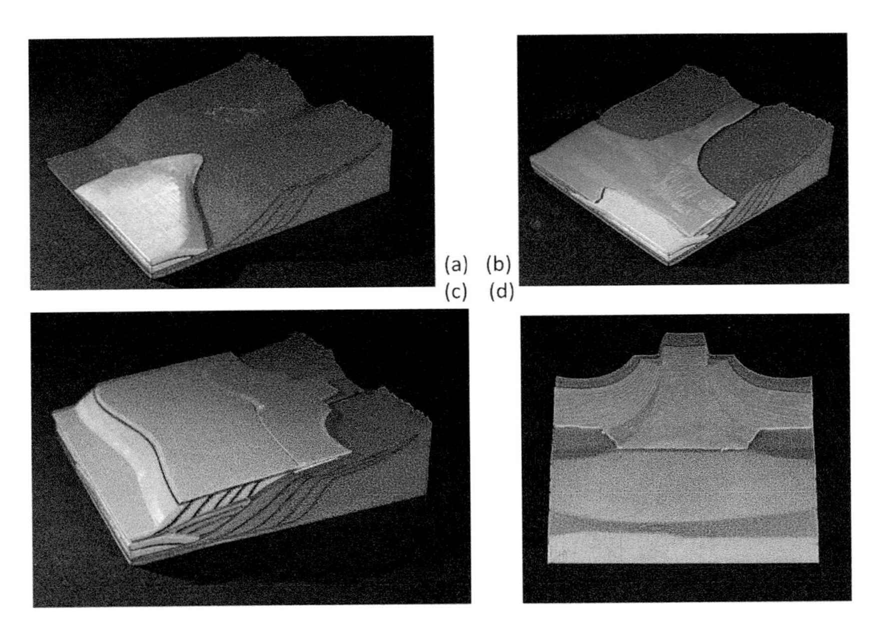

Figure H Sequential stratigraphy model taken apart: (a) high stand system tract in red and underwater cone in purple; (b) addition of a low stand system tract in blue; (c) a transgressive system tract in green (present but not very visible on the upper part of the margin) and a further high stand system tract in orange completing the cycle; (d) the two top sequences are turned over for better visualization (the two pieces are glued together). It will be noticed that during deposition of the transgressive material on the proximal part of the margin (the platform), sedimentation is much reduced in its distal part (not shown).

Source: Photo © Alexis Chézière

Figure I The seven crystal systems, from left to right and from bottom to top: monoclinic, hexagonal, orthorhombic, cubic (center), rhombohedral (white), triclinic and tetragonal.

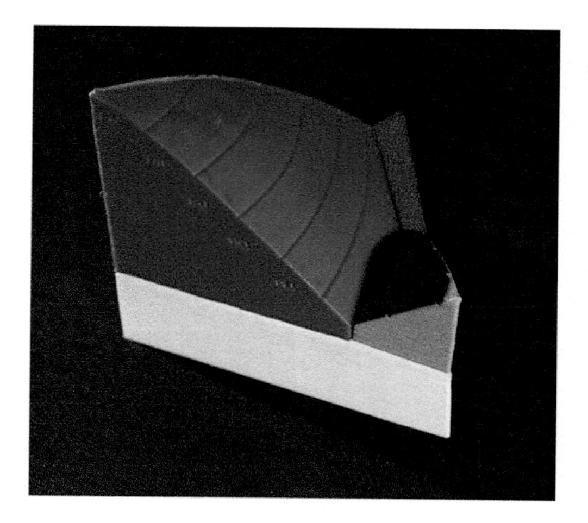

Figure J A 3D-printed model representing the forsterite (blue)-anorthite (green)-diopside (red) ternary diagram; these three phases are present in the rocks of the mantle (peridotite).

This model is technically difficult to produce because some parts do not have a flat surface (green and blue parts in Fig. H) that can be used as a basis for printing. To make this happen, we had to imagine a system of supports which were subsequently removed. The atelier is particularly useful in such cases.

Another field where the use of printed models seems obvious is mineralogy and petrography. It is known that crystalline solid matter is organized into seven crystal systems according to their morphological symmetry and the symmetry of their physical properties (Fig. I). This type of model, unlike the previous one, is very easy to create.

Finally, we present a model that is neither an idealized model of a geological object nor a simplification of an actual structure. This is a phase diagram (Fig. J) used in petrography and more generally in thermodynamics. So-called ternary diagrams (with three components) are often very difficult to visualize, hence the usefulness of a 3D physical model. The triangular base represents the composition in the form of mineralogical poles, the vertical dimension represents temperature. The model can be used to follow the melting or crystallization of a mixture of any composition. Thus, in crystallization mode (progressive decrease in temperature), a first mineral will crystallize; the composition of the residual liquid will gradually change (through removal of one of the phases) following the slope to the valley (cotectic line). Along the valley, two phases will be able to crystallize simultaneously until they reach the point where the valleys meet (ternary eutectic point) where the three phases crystallize together.

Glossary

Allochthon package of rocks (i.e., thrust sheet) that has been transported on a thrust fault or a detachment fault (in this case, the term "extensional allochthon" is used) that separates it from the autochthon.

Anticline arched structure in which the convexity is upward. Opposite: syncline.

Autochthon massif of rocks that was formed or produced in the location where it is now found. Opposite: allochthon.

Axial plane plane that divides the fold symmetrically. The axial plane can be underlined by a slaty cleavage.

Azimuth direction of a layer in relation to the north (*see* Strike).

Basalt a magmatic rock formed when magma reaches the earth's surface and cools quickly. This fine-grained rock is composed predominantly of silicate minerals (olivine, pyroxene, plagioclase) and with lesser amounts of oxides. Because of quick cooling, the size of minerals is small.

Basement term designing the resistant (generally crystalline or metamorphic) assemblage of rocks underlying the non-deformed or slightly deformed sedimentary rocks forming the cover. A major unconformity separates basement and cover.

Box fold fold in which the flat top of an anticline and/or the flat bottom of a syncline are bordered by vertical limbs.

Break-back sequence of thrusting sequence characterized by the development of structures from the foreland to the hinterland of a mountain belt. Such a sequence is very rare. Opposite: forward breaking sequence.

Brittle when a rock breaks, such a behavior is called brittle deformation. Opposite: ductile.

Calcarenite a sedimentary rock predominantly composed of carbonate grains.

Caledonian orogeny orogeny comprising events that occurred during the Lower Paleozoic. This orogenic system developed in the north of Europe and along the Canadian eastern coast.

Clay mineral that is hydrous aluminum phyllosilicate. By extension, a rock mostly composed of phyllosilicate minerals.

Cleavage planar feature that develops in fine grained rocks because of a pressure-solution process. In pure shear regime, a cleavage develops à right angle to the shortening direction and contains the stretching direction. A cleavage can develop at a late stage of deformation (after folding); in this case, it is parallel to the axial plane of the folds. It can also develop early (before folding) thanks to "layer parallel shortening" (LPS).

Collapse fall due to gravitational forces (e.g., the collapse of an orogenic system).

Collision orogen orogen resulting from the collision between two continents subsequently sutured together. Oceanic remnants (ophiolite) can be found along this suture, which characterizes the limit between both former continents.

Concentric fold a fold in which the thickness of each layer, measured perpendicular to the initial layering, is maintained all along the fold.

Conjugate faults two families of faults striking parallel but oppositely dipping. The two faults develop under the same stress field. The small bisector angle corresponds to the maximum stress axis σ_1.

Coupling in geology, two bodies are coupled when they mechanically respond identically.

Cover generic term designing sedimentary rocks deposited above a basement composed of old magmatic and magmatic rocks. A regional unconformity underlines the contact basement cover.

Craton an old and stable part of the continental crust since the beginning of the Neoproterozoic. Example: the West African craton.

Crust upper part of the lithosphere; the continental crust has a granitic mean composition; the oceanic crust has a basaltic mean composition.

Cutoff Intersection between a thrust surface and a stratigraphic layer. *See* Footwall cutoff and Hanging wall cutoff.

Décollement level weak layer with a ductile behavior allowing faults to follow it. A décollement level permits a decoupling of deformation on both side of it. A flat develops within a décollement level.

Detachment (décollement) weak layer that can be activated as a fault parallel to bedding; a thick detachment layer is required to allow the development of a detachment fold.

Detachment fault low angle fault generally convex upward allowing the lower crust and/or the mantle to be exhumed.

Diapir generic term designating a ductile mass penetrating overlying brittle rocks.

Dip dip and strike refer to the attitude of a geological plane (bedding, fault, joint, etc.). The dip gives the steepest angle of the slope of a geological surface by reference to a horizontal plane. A dip is given between 0° and 90°. *See also* Strike.

Ductility describes a rock behavior in which a solid material stretches under tensile stress.

Duplex stack of "horses" limited by thrusts that branch on a floor thrust below and merge with a roof thrust above.

Evaporites sedimentary rocks resulting from the concentration of an aqueous solution by evaporation and allowing the crystallization of minerals like halite, gypsum, etc.

External zones the less deformed zones of an orogenic system, located between the foreland and the internal zones. Typically, they do not record metamorphism and display fold-and-thrust belts.

Extrados external curve of a fold hinge. Opposite: intrados.

Fault surface delimiting two blocks with relative displacement between them.

Fault-bend fold a fault-bend fold is a fold developed by bending over a staircase thrust fault (flat-ramp-flat).

Fault-propagation fold fold developing when a propagating thrust fault terminates upsection by transferring its shortening to a fold developing at its tip.

Flat fault surface parallel to the strata; a flat corresponds to a décollement level and develops in weak rocks (shale, marls, etc.) and connect two ramps.

Flexural basin basin formed primarily because of the downward flexing of the lithosphere in response to the weight of a mass of rocks, for instance an adjacent mountains belt. Therefore, such a basin can develop at the front of mountains (e.g., the Swiss Molassic basin at the front of the Western Alps).

Flexural slip slipping of sedimentary strata along bedding planes during folding.

Fold-and-thrust belt series of mountainous foothills adjacent to an orogenic belt, in which a close association of folds and thrust faults records deformation.

Footwall cutoff intersection between a thrust surface and a stratigraphic layer in the footwall of a thrust fault.

Forward breaking sequence sequence characterized by the development of structures from the hinterland to the foreland of a mountain belt. Opposite: break-back sequence.

Frontal imbrication lengthening of a thrust wedge characterized the development of a new thrust in the frontal part of the developing wedge.

Gabbro mafic intrusive igneous rock with a slow cooling history allowing the development of coarse-grained texture. Gabbro is the result of the partial melting of peridotites and is one of the main components of the oceanic crust. They mainly contain silicates (pyroxene + amphibole +olivine + plagioclase) and some oxides.

Graben topographic depression bounded by oppositely dipping normal faults. Opposite: horst.

Granite intrusive igneous rock with a slow cooling history allowing the development of coarse-grained texture. Granites are the result of the partial melting of continental crust or the result of differentiation of partial melting of the mantle. Granites mainly contain silicates (quartz + K-feldspar + plagioclase + mica).

Growth strata syntectonic deposits (i.e., deposited during deformation). Therefore, their age defines the timing of deformation. They can be found in extensional or compressional settings.

Halokinetic sequence specific syn-sedimentary sequence recording the timing and the style of salt activity.

Hanging wall cutoff intersection between a thrust surface and a stratigraphic layer in the hanging wall of a thrust fault.

Hiatus discontinuity in the age of strata link to a non-deposition of sediments or to an erosional event. An unconformity is always associated with a hiatus, but the reverse is not true.

Hinge part of the fold where the curvature is greatest. In a fold, hinges are linked by limbs.

Horse package of rocks bounded by faults.

Horst topographic high bounded by two oppositely dipping normal faults; the raised block is the footwall of the normal faults. Opposite: graben.

Inlier (or exhumed anticlinal inlier) area of older rocks surrounded by younger rocks. Inliers result from the erosion of the overlying younger rocks revealing a restricted exposure of the older rocks. Faulting or folding generally promote the observed outcrop pattern. Example: the Pays de Bray inlier in the Paris basin.

Internal zones zones located in the core of orogenic systems. Typically, they display metamorphism and magmatic rocks and contain the suture, when it exists.

Intracontinental orogen orogenic system developed within a continent on the site of a pre-existing failed rift (i.e., a rift system, which did not give birth to an oceanic domain).

Intrados the interior curve of a fold hinge. Opposite: extrados.

Joint cracks or fractures along which there is no visible movement.

Kinematics study of displacement and, consequently, in geology, study of successive geometries through time. A kinematic model describes the evolution of an initial state to a final state without physical or mechanical concepts.

Kink band a band of parallel layers comprises between two kink planes (i.e., the axial plane of a kink fold).

Kink fold a type of fold with a sharp angular hinge and straight limb. Kink develops in foliated or layered rocks. Some authors consider that term kink fold must be used only in schists and prefer the term "chevron fold" for sedimentary layered rocks.

Kink method method using a simplified geometry to construct fold in which each fold is represented as a kink fold. John Suppe's equations, allowing the construction of balanced cross-sections, follow this methodology.

Kink plane axial plane of a kink fold.

Klippe remnant of a thrust nappe surrounded, on map view, by a thrust fault and isolated from the rest of the thrust nappe by erosion.

Left lateral (or sinistral) a left-lateral strike-slip fault is a fault where the left block moves toward you and the right block moves away when you stay on the fault and look at it. Opposite: right-lateral (dextral).

Limb in a fold, the limbs correspond to the less curved part of the fold between two hinges.

Limestone sedimentary rock mostly composed of calcium carbonate.

Listric fault fault with a concave upward geometry, the dip of which decreases in depth until merging a décollement level.

Lithosphere earth's lithosphere is the external envelop of the planet. It is made of the crust (oceanic or continental) and of the upper part of the mantle.

Localization of the deformation process occurring when deformation concentrates in a restricted area (along a fault, a hinge, etc.). Opposite: distributed deformation.

Mantle the earth's mantle is the biggest layer inside earth. It is bounded by the core and the crust and made of silicate rocks mainly composed of olivine and pyroxene. The upper part of the mantle is included in the lithosphere.

Margin transition zone between a continental domain and an oceanic domain.

Old massif generic term designing a zone where the basement is cropping out (*see also* Shield).

Ophiolite section of an oceanic crust that has been abducted and transported over the continental crust during an orogeny. Like the oceanic crust ophiolite is made up of an association of peridotite, gabbro and basalt. In an orogen, ophiolite underlies the suture.

Orogen (orogenic system) generic term designing a young or ancient system of mountain belt.

Orogenic cycle cycle describing the opening (rifting), then the closure, of an oceanic domain leading to the development of a mountain belt; the orogenic cycle is also called the Wilson cycle from the Canadian geologist Tuzo Wilson (1908–1993), father of the concept.

Out-of-sequence thrust thrusts that do not follow a classical order of propagation. They generally are isolated thrusts developed at the back of the thrust front and cutting through an existing chain of structures. They exist in almost all mountain belts.

Peridotite peridotite is a generic name used for coarse-grained, ultramafic igneous rock. Peridotites usually contain olivine and pyroxene as primary minerals. Peridotite is the main component of the Upper Mantle.

Plane strain deformation two-dimensional state of strain in which all the shape changes happen on a single plane; by contrast, nothing happens in the direction normal to this plane.

Plutonic rocks intrusive rocks resulting from the solidification of magma at depth. In plutonic rocks, the crystals can be seen with the naked eye (granular texture).

Pop-up structure uplift between a fore thrust and a back thrust. A pop-up can develop during the late stages of detachment folding (box folding in particular).

Ramp a fault cutting through the strata and joining two flats; fault-bend folds and fault-propagation folds develop over a ramp.

Ramp anticline generic term describing a fold developed over a ramp.

Regression geological process resulting from a sea level fall. The consequence is that submerged areas become exposed above sea level. Opposite: transgression.

Rheology study of matter behavior in a given stress configuration. It describes how a medium, a rock in particular, behaves when a force is applied to it.

Rift narrow zone where the lithosphere is being thinned and pulled apart. The net result is the development of a depression bounded by extensional faults, also called a rift or graben.

Right lateral (or dextral) a right-lateral strike-slip fault is a fault where the right block moves toward you and the left block moves away when you stay on the fault and look at it. Opposite: left-lateral (sinistral).

Roll-over anticline syn-depositional structure developed in the hanging wall of an extensional listric fault.

Salt canopy large volume of salt resulting from the coalescence of a number of salt glaciers.

Salt glacier flow of salt, which forms when a rising diapir reaches the surface of the earth and floods along this surface.

Salt plug (or salt stock) kind of salt diapir presenting a subcircular shape in map view.

Salt wall (or salt ridge) kind of diapir presenting an elongated shape in map view.

Sandstone clastic sedimentary rock composed of sand-sized grains (quartz, feldspar, lithic with variable quantity) and a muddy matrix of variable quantity.

Sequence of thrusting order in which the thrust-faults develop in a fold-thrust belt.

Shield a large stable area where the basement composed of magmatic and/or metamorphic rocks outcrops, e.g., Baltic shield.

Strata (plural of stratum) parallel layers formed at the earth's surface. Strata can be sedimentary or not.

Stratigraphic contact contact due to sedimentary deposit of a given rock over an older one. It is opposed to a tectonic contact, which is a faulted contact. An unconformity is a particular stratigraphic contact.

Stress (normal or tangential) in mechanics, a stress is a force per unit area. Consequently, a stress is like a pressure, and the unit is Pascal (Pa). The difference is that pressure always acts perpandicular to the surface, while stress acts either perpandicular or parallel to the surface.

Stress tensor in continuum mechanics, a stress tensor consists of nine components σ_{ij} that completely define the state of stress at a point inside a deformed material.

Strike *see* azimuth.

Stylolite irregular surfaces (with sharp peaks) developed mostly in carbonated rocks and resulting from pressure-solution process. Insoluble minerals or organic matter underlies the stylolites.

Subsidence gradual downward movement of the earth's surface with slight or no horizontal displacement. The causes of subsidence can be tectonic (thinning or flexing of the lithosphere) or thermal. The consequence of subsidence is the development of an accommodation space available for the deposition of sedimentary rocks. The development of a sedimentary basin results from subsidence. Opposite: uplift.

Subsurface domain constituted by the rocks located near (few kilometers) but not exposed at the surface of the earth.

Suture trace of an ancient oceanic domain in a mountain belt.

Syncline arched structure in which the convexity is downwards. Opposite: anticline.

Tectonic inversion reactivation of a preexisting fault with a different movement than the first one.

Tectonic regime context defined by one of the three principal stress configurations, namely the tensional (or extensional), the compressional and the strike-slip regimes.

Tectonic window outcrop of autochthonous rocks located below a thrust sheet. On map view, a tectonic window is bounded by a thrust fault. It is worth noting that a tectonic window frequently forms a culmination (tectonic stack) piercing the overlying thrust sheet.

Tension gash a fracture characterized by an opening perpendicular to its walls. Tension gashes are filled by recrystallized minerals (calcite, quartz, etc.). This category of fractures develops in presence of fluids when rocks present low tensile strength.

Thick skinned deformation deformation in which the basement is involved. Opposite: thin-skinned deformation.

Thin-skinned deformation deformation in which the basement is not involved. Opposite: thick-skinned deformation.

Thrust fault reverse fault with a low or very low dip. A thrust fault floors a thrust sheet.

Thrust sheet rock package transported for several kilometers along a thrust fault.

Transgression geological process resulting from a sea level rise. The consequence is that outcropping areas become submerged below sea level. Opposite: regression.

Unconformity old erosional surface separating two packages of rocks of different ages, the older being deformed and truncated by the erosional surface, the younger being parallel to it.

Under-thrusting tectonic process during which the rocks located in the footwall of a thrust fault move under the relatively static rocks of the hanging wall. This term is often misunderstood because thrusting is a relative movement. Opposite: over-thrusting.

Variscan (or Hercynian) orogeny old collisional orogenic system developed during the Upper Paleozoic. The Variscan orogeny is exposed in Western and Southern Europe, in North America (Appalachian mountains) and in West Africa along the Central Atlantic in Morocco and Mauritania. The Pangea mega-continent is the direct consequence of the Variscan orogeny.

Vergence overall direction of over-thrusting in a fold-and-thrust belt. This term, sometime misunderstood, can also by applied at the scale of a single structure (fault, fold).

Stratigraphic chart

Eonothem / Eon	Erathem / Era	System / Period	Series / Epoch	Stage / Age	GSSP	numerical age (Ma)
						present
Phanerozoic	Cenozoic	Quaternary	Holocene		✓	0.0117
			Pleistocene	Upper		0.126
				Middle		0.781
				Calabrian	✓	1.806
				Gelasian	✓	2.588
		Neogene	Pliocene	Piacenzian	✓	3.600
				Zanclean	✓	5.333
			Miocene	Messinian	✓	7.246
				Tortonian	✓	11.62
				Serravallian	✓	13.82
				Langhian		15.97
				Burdigalian		20.44
				Aquitanian	✓	23.03
		Paleogene	Oligocene	Chattian		28.1
				Rupelian	✓	33.9
			Eocene	Priabonian		38.0
				Bartonian		41.3
				Lutetian	✓	47.8
				Ypresian	✓	56.0
			Paleocene	Thanetian	✓	59.2
				Selandian	✓	61.6
				Danian	✓	66.0
	Mesozoic	Cretaceous	Upper	Maastrichtian	✓	72.1 ±0.2
				Campanian		83.6 ±0.2
				Santonian		86.3 ±0.5
				Coniacian		89.8 ±0.3
				Turonian	✓	93.9
				Cenomanian	✓	100.5
			Lower	Albian		~ 113.0
				Aptian		~ 125.0
				Barremian		~ 129.4
				Hauterivian		~ 132.9
				Valanginian		~ 139.8
				Berriasian		~ 145.0

Eonothem / Eon	Erathem / Era	System / Period	Subsystem / Subperiod	Series / Epoch	Stage / Age	GSSP	numerical age (Ma)
							145.0 ± 0.8
Phanerozoic	Mesozoic	Jurassic		Upper	Tithonian		152.1 ±0.9
					Kimmeridgian		157.3 ±1.0
					Oxfordian		163.5 ±1.0
				Middle	Callovian		166.1 ±1.2
					Bathonian	✓	168.3 ±1.3
					Bajocian	✓	170.3 ±1.4
					Aalenian	✓	174.1 ±1.0
				Lower	Toarcian		182.7 ±0.7
					Pliensbachian	✓	190.8 ±1.0
					Sinemurian	✓	199.3 ±0.3
					Hettangian	✓	201.3 ±0.2
		Triassic		Upper	Rhaetian		~ 208.5
					Norian		~ 228
					Carnian	✓	~ 235
				Middle	Ladinian	✓	~ 242
					Anisian		247.2
				Lower	Olenekian		251.2
					Induan	✓	252.6
	Paleozoic	Permian		Lopingian	Changhsingian	✓	254.2 ±0.1
					Wuchiapingian	✓	259.9 ±0.4
				Guadalupian	Capitanian	✓	265.1 ±0.4
					Wordian	✓	268.8 ±0.5
					Roadian	✓	272.3 ±0.5
				Cisuralian	Kungurian		279.3 ±0.6
					Artinskian		290.1 ±0.1
					Sakmarian		295.5 ±0.4
					Asselian	✓	298.9 ±0.2
		Carboniferous	Pennsylvanian	Upper	Gzhelian		303.7 ±0.1
					Kasimovian		307.0 ±0.1
				Middle	Moscovian		315.2 ±0.2
				Lower	Bashkirian	✓	323.2 ±0.4
			Mississippian	Upper	Serpukhovian		330.9 ±0.2
				Middle	Visean	✓	346.7 ±0.4
				Lower	Tournaisian	✓	358.9 ±0.4

Eonothem / Eon	Erathem / Era	System / Period	Series / Epoch	Stage / Age	GSSP	numerical age (Ma)
						358.9 ± 0.4
Phanerozoic	Paleozoic	Devonian	Upper	Famennian	✓	372.2 ±1.6
				Frasnian	✓	382.7 ±1.6
			Middle	Givetian	✓	387.7 ±0.8
				Eifelian	✓	393.3 ±1.2
			Lower	Emsian	✓	407.6 ±2.6
				Pragian	✓	410.8 ±2.8
				Lochkovian	✓	419.2 ±3.2
		Silurian	Pridoli		✓	423.0 ±2.3
			Ludlow	Ludfordian	✓	425.6 ±0.9
				Gorstian	✓	427.4 ±0.5
			Wenlock	Homerian	✓	430.5 ±0.7
				Sheinwoodian	✓	433.4 ±0.8
			Llandovery	Telychian	✓	438.5 ±1.1
				Aeronian	✓	440.8 ±1.2
				Rhuddanian	✓	443.4 ±1.5
		Ordovician	Upper	Hirnantian	✓	445.2 ±1.4
				Katian	✓	453.0 ±0.7
				Sandbian	✓	458.4 ±0.9
			Middle	Darriwilian	✓	467.3 ±1.1
				Dapingian	✓	470.0 ±1.4
			Lower	Floian	✓	477.7 ±1.4
				Tremadocian	✓	485.4 ±1.9
		Cambrian	Furongian	*Stage 10*		~ 489.5
				Jiangshanian	✓	~ 494
				Paibian	✓	~ 497
			Series 3	Guzhangian	✓	~ 500.5
				Drumian	✓	~ 504.5
				Stage 5		~ 509
			Series 2	*Stage 4*		~ 514
				Stage 3		~ 521
			Terreneuvian	*Stage 2*		~ 529
				Fortunian	✓	541.0 ±1.0

Eonothem / Eon	Erathem / Era	System / Period	GSSP GSSA	numerical age (Ma)
				~ 541
Precambrian	Proterozoic	Neo-proterozoic	Ediacaran	GSSP	~ 635
			Cryogenian	GSSA	850
			Tonian	GSSA	1000
		Meso-proterozoic	Stenian	GSSA	1200
			Ectasian	GSSA	1400
			Calymmian	GSSA	1600
		Paleo-proterozoic	Statherian	GSSA	1800
			Orosirian	GSSA	2050
			Rhyacian	GSSA	2300
			Siderian	GSSA	2500
	Archean	Neo-archean		GSSA	2800
		Meso-archean		GSSA	3200
		Paleo-archean		GSSA	3600
		Eo-archean		GSSA	4000
	Hadean				~ 4600

Units of all rank are in the process of being defined by Global Boundary Stratotype Section and Points (GSSP) for their lower boundaries, including those of the Archean and Proterozoic, long defined by Global Standard Stratigraphic Ages (GSSA). Charts and detailed information on ratified GSSPs are available at the website **http://www.stratigraphy.org**

Numerical ages are subject to revision and do not define units in the Phanerozoic and the Ediacaran; only GSSPs do. For boundaries in the Phanerozoic without ratified GSSPs or without constrained numerical ages, an approximate numerical age (~) is provided.

Numerical ages for all systems except Triassic, Cretaceous and Precambrian are taken from '*A Geologic Time Scale 2012*' by Gradstein et al. (2012); those for the Triassic and Cretaceous were provided by the relevant ICS subcommissions.

Coloring follows the Commission for the Geological Map of the World. http://www.cgm.org

Chart drafted by K.M. Cohen, S. Finney, P.L. Gibbard

Bibliography

This list includes the documents cited in the text, in particular PhD theses and one HDR (Habilitation for Research Supervision) thesis defended in Cergy-Pontoise University (now CY Cergy Paris Université), as well as articles and books that we recommend for further reading on some questions addressed in the book.

Agard, P. & Lemoine, M. (2003) Visages des Alpes: structure et évolution dynamique. CCGM-CGMW.

Biot, M.A. (1961) Theory of folding of stratified viscoelastic media and its implications in tectonics and orogenesis. *Geological Society of America Bulletin*, 72(11), 1595–1620.

Boillot, G., Huchon, P., Lagabrielle, Y. & Boutler, J. (2013) Introduction à la géologie, la dynamique de la Terre. Dunod.

Boyer, S.E. & Elliot, D. (1982) Thrust systems. *American Association Petroleum Geology Bulletin*, 66, 1196–1230.

Brun, J.P. & Fort, X. (2008) Entre Sel et Terre, structure et mécanisme de la tectonique salifère. Société Géologique de France & Vuibert. 153 pp.

Brun, J.P. & Fort, X. (2011) Salt tectonics at passive margins: Geology versus model. *Marine and Petroleum Geology*, 28, 1123–1145.

Buil, D. (2002) L'approche cinématique du plissement naturel: intérêts et limites. Développements autour de la notion de "trishear." PhD Thesis, Université de Cergy-Pontoise, France.

Butler, R.W.H. (1982) The terminology of structures in thrust belts. *Journal Geophysical Research*, 4, 239–246.

Butler, R.W.H., Bond, C.E., Cooper, M.A. & Watkins, H. (2018) Interpreting structural geometry in fold-thrust belts: Why style matters. *Journal of Structural Geology*, 114, 251–273.

Coirier, J. & Nadot-Martin, C. (2013) Mécanique des milieux continus, cours et exercices corrigés, Collection Sciences Sup. Dunod.

Cordary, D. (1995) Mécanique des sols. Tec. et Doc. Lavoisier.

Cubas, N. (2009) Séquences de chevauchement, prédictions mécaniques, validation analogique et application à la chaîne de l'Agrio (Argentine), PhD Thesis, University Paris-Sud (Orsay), 233 pp.

Dahlstrom, C.D.A. (1969) Balanced cross sections. *Canadian Journal of Earth Sciences*, 6, 743–757.

Dahlstrom, C.D.A. (1970) Structural geology in the eastern margin of the Canadian Rocky Mountains. *Bulletin of Canadian Petroleum Geology*, 18, 332–406.

Debelmas, J., Mascle, G. & Basile, C. (2008) Les grands structures géologiques, 5ème édition. Dunod.

Deleuze, G. (1988) The Fold, Leibniz and the Baroque (le pli, Leibniz et le Baroque). Les Editions de Minuit, Paris, 192 pp.

De Sitter, L.U. (1956) Structural Geology. McGraw-Hill.

Ellenberger, F. (1967) Les interférences de l'érosion et de la tectonique tangentielle tertiaire dans le Bas-Languedoc (principalement dans l'arc de Saint-Chinian), notes sur les charriages cisaillants. *Revue de Géographie physique et Géologie dynamique*, 9, 87–143.

Elliott, D. (1976) The energy balance and deformation mechanisms of thrust sheets. *Philosophical Transactions Royal Society of London*, (A), 283, 289–312.

Fossen, H. (2016) Structural Geology, 2nd edition. Cambridge University Press.

Freud S. (1920) A general introduction to Psychoanalysis, Translation in 2012 by Stanley Hall, Wordsworth classics of World litterature, eighteenth lecture, Hestfordshire.

Frizon de Lamotte, D. & Buil, D. (2002) La question des relations entre failles et plis dans les zones externes des chaînes demontagnes. Ebauche d'une histoire des idées au cours du XXe siècle. Travaux du Comité Français d'Histoire de la Géologie – Troisième série -T.XVI.

Frizon de Lamotte, D. *et al.* (2013) Evidence for Late Devonian vertical movements and extensional deformation in northern Africa and Arabia: Integration in the geodynamics of the Devonian world. *Tectonics*. doi: 10.1029/2012TC003133.

Gallup, W. (1951) Geology of Turner valley oil and gas field, Alberta, Canada. *American Association Petroleum Geology Bulletin*, 34, 797–821.

Geikie, J. (1912). Structural and Field Geology, 3rd edition. Oliver and Boyd, Edinburgh, 452 pp.

Gély, J.P. & Hanot, F. *et al.* (2014) Le Bassin Parisien, un nouveau regard sur la géologie. Association des Géologues du Bassin de Paris.

Giles, K.A. & Rowan, M.G. (2012) Concepts in halokinetic sequence deformation and stratigraphy: From Asolp *et al.* Salt tectonics, sediments and prospectivity. *Geological Society of London, Special Publication*, 363, 7–31.

Gimeno-Vives, O., Mohn, G., Bosse, V., Haissen, F., Zaghloul, M.N., Atouabat, A. & Frizon de Lamotte, D. (2019) The Mesozoic margin of the Maghrebian Tethys in the Rif belt (Morocco): Evidence for polyphaser rifting and related magmatic activity. *Tectonis*, 38. doi.org/10.1029TCOàà5508.

Goguel, J. (1952) Traité de Tectonique. Masson.

Gracq, J. (1986) Reading while writing (En lisant en écrivant), José Corti, Paris, 302 pp.

Graveleau, F., Malavieille, J. & Dominguez, S. (2012) Experimental modelling of orogenic wedges: A review. *Tectonophysics*, 538–540, 1–66. doi: 10.1016/j.tecto.2012.01.027.

Hall, J. (1815) On the vertical position and convolutions of certain strata and their relationship with granite. *Transactions of the Royal Society of Edinburgh*, 7, 79–108.

Hubbert, K. (1951) Mechanical basis for certain familiar geologic structures. *Geological Society of America Bulletin*, 62, 355–372.

Hutton, J. (1795) Theory of the Earth with proofs and illustrations, 2 vol. Edinburgh, Creech.

Jolivet, L. (1997) La déformation des continents, exemples régionaux. Hermann.

Jolivet, L. & Nataf, H.C. (1998) Géodynamique. Dunod.

Manatschal, G. *et al.* (2001) The role of detachment faulting in the formation of an ocean: Continent transition: Insights from the Iberia Abyssal Plain. *Geological Society, London, Special Publications*, 187, 405–428.

Mattauer, M. (1999a) Ce que disent les pierres. Bibliothèque pour la Science.

Mattauer, M. (1999b) Monts et Merveilles, beautés et richesses de la géologie. Hermann.

Mercier, E. (1996) Les plis de propagation de rampe: cinématique, modélisation et importance dans la tectogenèse, Mémoire d'Habilitation à Diriger des Recherches (HDR). Université de Cergy-Pontoise, France.

Mercier, J., Vergély, P. & Missenard, Y. (2011) Tectonique, 3ème édition. Dunod.

Molinaro, M. (2004) Geometry and Kinematics of the SE Zagros Mountains (Iran): Structural Geology and Geophysical Modelling in a Young Collisional Fold-Thrust Belt. PhD Thesis, Université de Cergy-Pontoise, France.

Price, N.J. (1966) Fault and Joint Development in Brittle and Semi-Brittle Rock. Pergamon Press.

Ramsay, J.G. & Huber, M.I. (1987) Modern Structural Geology, Volume 1, Strain Analysis, Volume 2, Folds and Fractures. Academic Press.
Renard, M., Lagabrielle, Y., Martin, E. & de Rafélis, M. (2018) Eléments de géologie, 16ème édition. Dunod.
Rich, J.L. (1934) Mechanics of low-angle overthrust faulting as illustrated by Cumberland Thrust Block, Virginia, Kentucky, and Tennessee. *American Association of Petroleum Geologists Bulletin*, 18, 1584–1596.
Risnes, R. (2001) Deformation and yield in high porosity outcrop chalk. *Physics and Chemistry of the Earth A*, 26, 53– 57.
Robert, C. & Bousquet, R. (2013) Géosciences, la dynamique du système terrestre. Belin.
Robert, R. (2018) Etude de la déformation dans une formation granulaire poreuse en régime compressif: du terrain au laboratoire. PhD Thesis, Université de Cergy-Pontoise, France.
Salençon, J. (2001) Mécanique des Milieux Continus, tome I: concepts généraux. Edition de l'Ecole Polytechnique.
Schreurs, G. et al. (2006) Analogue Benchmarks of Shortening and Extension Experiments, Special Publication, 253. Geological Society of London.
Sherkati, S., Molinaro, M., Frizon de Lamotte, D. & Letouzey, J. (2005) Detachment folding in the Central and Eastern Zagros fold-belt (Iran): Salt mobility, multiple detachments and late basement control. *Journal of Structural Geology*, 27, 1680–1696.
Sorel, D. & Vergely, P. (2018) Atlas d'initiation aux cartes et coupes géologiques, 4ème édition. Dunod.
Souque, C. (2002) Magnétisme Structural dans les chaînes de chevauchement-plissement: développements analytiques et exemples d'utilisation dans les Corbières. PhD Thesis, Université de Cergy-Pontoise, France.
Suppe, J. (1983) Geometry and kinematics of fault-bend folding. *American Journal of Sciences*, 283, 684–721.
Suppe, J. (1985) Principles of Structural Geology. Prentice-Hall Inc.
Tari, G. & Jabour, H. (2013) Salt tectonics along the Atlantic margin of Morocco. *Geological Society of London*, Special Publication. doi: 10.1144/SP369.23.
Teixell, A. *et al.* (2003) Tectonic shortening and topography in the Central High Atlas (Morocco). *Tectonics*. doi: 10.1029/2002TC1460.
Turcotte, D.L. & Schubert, G. (1982) Geodynamics, Application of Continuous Physics to Geological Problems. J. Wiley and Sons.
Twiss, R.J. & Moores, E.M. (2007) Structural Geology, 2nd edition. Freeman.
Wrobel-Daveau, J.-C. (2011) From Rifting to Current Collision, Vertical Movements and Propagation of the Deformation in the Zagros Belt, Iran: Insights from Section Balancing and Detrital Low Temperature Thermochronology. PhD Thesis, Université de Cergy-Pontoise, France.

Index

Note: Page numbers in **bold** refer to entries in the glossary.